Iterative Optimization in Inverse Problems

MONOGRAPHS AND RESEARCH NOTES IN MATHEMATICS

Published Titles

Iterative Optimization in Inverse Problems, Charles L. Byrne

Forthcoming Titles

Modeling and Inverse Problems in the Presence of Uncertainty, H. Thomas Banks, Shuhua Hu, and William Clayton Thompson

Sinusoids: Theory and Technological Applications, Prem K. Kythe

Stochastic Cauchy Problems in Infinite Dimensions: Generalized and Regularized Solutions, Irina V. Melnikova and Alexei Filinkov

Signal Processing: A Mathematical Approach, Charles L. Byrne

Monomial Algebra, Second Edition, Rafael Villarreal

Groups, Designs, and Linear Algebra, Donald L. Kreher

Geometric Modeling and Mesh Generation from Scanned Images, Yongjie Zhang

Difference Equations: Theory, Applications and Advanced Topics, Third Edition, Ronald E. Mickens

Set Theoretical Aspects of Real Analysis, Alexander Kharazishvili

Method of Moments in Electromagnetics, Second Edition, Walton C. Gibson

The Separable Galois Theory of Commutative Rings, Second Edition, Andy R. Magid

Dictionary of Inequalities, Second Edition, Peter Bullen

Actions and Invariants of Algebraic Groups, Second Edition, Walter Ferrer Santos and Alvaro Rittatore

Practical Guide to Geometric Regulation for Distributed Parameter Systems, Eugenio Aulisa and David S. Gilliam

Analytical Methods for Kolmogorov Equations, Second Edition, Luca Lorenzi

Handbook of the Tutte Polynomial, Joanna Anthony Ellis-Monaghan and Iain Moffat

MONOGRAPHS AND RESEARCH NOTES IN MATHEMATICS

Iterative Optimization in Inverse Problems

Charles L. Byrne
University of Massachusetts
Lowell, USA

CRC Press
Taylor & Francis Group
Boca Raton London New York

CRC Press is an imprint of the
Taylor & Francis Group, an **informa** business
A CHAPMAN & HALL BOOK

CRC Press
Taylor & Francis Group
6000 Broken Sound Parkway NW, Suite 300
Boca Raton, FL 33487-2742

ISBN 13: 978-1-4822-2233-3 (hbk)

Library of Congress Cataloging-in-Publication Data

Byrne, Charles L., 1947- author.
Iterative optimization in inverse problems / Charles L. Byrne, University of Massachusetts, Lowell.
pages cm. -- (Monographs and research notes in mathematics)
Includes bibliographical references and index.
ISBN 978-1-4822-2233-3 (hardback)
1. Iterative methods (Mathematics) 2. Algorithms. 3. Inverse problems (Differential equations) I. Title.

QA297.8.B974 2008
515'.357--dc23 2013038279

Visit the Taylor & Francis Web site at
http://www.taylorandfrancis.com

and the CRC Press Web site at
http://www.crcpress.com

To Eileen,
my wife for forty three years.

Contents

Preface

It is not easy to give a precise definition of an inverse problem, but, as they often say about other things, we know one when we see one. Loosely speaking, direct problems involve determining the effects of known causes. What will be the temperatures later at points within the room, given the current temperatures? Indirect, or inverse, problems go the other way, as we attempt to infer causes from observed effects. What was the temperature distribution in the room initially, given that we have measured the temperatures at several later times? Most remote-sensing problems are inverse problems.

Magnetic resonance, acoustic, and optical remote-sensing problems typically involve measuring Fourier transform values of the function we wish to estimate. Transmission and emission tomographic image reconstruction is often described this way, as well. In Chapter 1 I describe a typical remote-sensing problem of this type, to illustrate the ways in which the measured data is often limited and to demonstrate how projection-based methods and minimum-norm approximate solutions can be employed. The algorithms used here are not usually iterative, although iterative techniques can be used to avoid difficult computational steps.

My exposure to iterative algorithms began with the *algebraic reconstruction technique* (ART) and the *expectation maximization maximum likelihood* (EMML) approaches to medical imaging, both methods described briefly in Chapter 1. Both the ART and the EMML algorithm are used for medical image reconstruction, but there the resemblance seemed to end. The ART is a sequential algorithm, using only one data value at a time, while the EMML is simultaneous, using all the data at each step. The EMML has its roots in statistical parameter estimation, while the ART is a deterministic method for solving systems of linear equations. The ART can be used to solve any system of linear equations, while the solutions sought using the EMML method must be nonnegative vectors. The ART employs orthogonal projection onto hyperplanes, while the EMML algorithm is best studied using the Kullback-Leibler, or cross-entropy, measure of distance. The ART converges relatively quickly, while the EMML is known to be slow. If there has been any theme to my work over the past decade, it is unification. I have tried to make connections among the various algorithms and problems I have studied. Connecting the ART and the EMML seemed like a good place to start.

The ART led me to its multiplicative cousin, the MART, while the EMML brought me to the simultaneous MART (SMART), showing that the statistical EMML could be viewed as an algorithm for solving certain systems of linear equations, thus closing the loop. There are block-iterative versions of all these algorithms, in which some, but not all, of the data is used at each step of the iteration. These tend to converge more quickly than their simultaneous relatives. Casting the EMML and SMART algorithms in terms of cross-entropic projections led to a computationally simpler variant of the MART, called the EMART. The Landweber and Cimmino algorithms are simultaneous versions of the ART. Replacing the cross-entropy distance with distances based on Fermi-Dirac entropy provided iterative reconstruction algorithms that incorporated upper and lower bounds on the pixel values. The next issue seemed to be how to connect these algorithms with a broader group of iterative optimization methods.

My efforts to find unification among iterative methods has led me recently to sequential optimization. A wide variety of iterative algorithms used for continuous optimization can be unified within the framework of sequential optimization. The objective in sequential optimization is to replace the original problem, which often is computationally difficult, with a sequence of simpler optimization problems. The most common approach is to optimize the sum of the objective function and an auxiliary function that changes at each step of the iteration. The hope is that the sequence of solutions of these simpler problems will converge to the solution of the original problem.

Sequential unconstrained minimization (SUM) methods [126] for constrained optimization are perhaps the best known sequential optimization methods. The auxiliary functions that are added are selected to enforce the constraints, as in barrier-function methods, or to penalize violations of the constraints, as in penalty-function methods.

We begin our discussion of iterative methods with auxiliary-function (AF) algorithms, a particular class of sequential optimization methods. In AF algorithms the auxiliary functions have special properties that serve to control the behavior of the sequence of minimizers. As originally formulated, barrier- and penalty-function methods are not AF algorithms, but both can be reformulated so as to be included in the AF class. Many other well-known iterative methods can also be shown to be AF methods, such as proximal minimization algorithms using Bregman distances, projected gradient descent, the CQ algorithm, the forward-backward splitting method, MART and SMART, EMART and the EMML algorithm, alternating minimization (AM), and majorization minimization (MM), or optimality transfer techniques, and the more general expectation maximization maximum likelihood EM algorithms in statistics. Most of these methods enjoy additional properties that serve to motivate the definition of the SUMMA class of algorithms, a useful subclass of AF methods.

Some AF algorithms can be described as fixed-point algorithms, in which the next vector in the sequence is obtained by applying a fixed operator to the previous vector and the solution is a fixed point of the operator. This leads us to our second broad area for discussion, iterative fixed-point methods. Operators that are nonexpansive in some norm provide the most natural place to begin discussing convergence of such algorithms. Being nonexpansive is not enough for convergence, generally, and we turn our attention to more restrictive classes of operators, such as the averaged and paracontractive operators.

Convexity plays an important role in optimization and a well developed theory of iterative optimization is available when the objective function is convex. The gradient of a differentiable convex function is a monotone operator, which suggests extending certain optimization problems to variational inequality problems (VIP) and modifying iterative optimization methods to solve these more general problems. Algorithms for VIP can then be used to find saddle points.

Our discussion of iterative methods begins naturally within the context of the Euclidean distance on finite-dimensional vectors, but is soon broadened to include other useful distance measures, such as the l^1 distance, and cross-entropy and other Bregman distances. Within the context of the Euclidean distance, orthogonal projection onto closed convex sets plays an important role in constrained optimization. When we move to other distances we shall attempt to discover the extent to which more general notions of projection can be successfully employed.

Problems in remote sensing, such as radar and sonar, x-ray transmission tomography, PET and SPECT emission tomography, and magnetic resonance imaging, involve solving large systems of linear equations, often subject to constraints on the variables. Because the systems involve measured data as well as simplified models of the sensing process, finding exact solutions, even when available, is usually not desirable. For that reason, iterative regularization algorithms that reduce sensitivity to noise and model error and produce approximate solutions of the constrained linear system are usually needed.

In a number of applications, such as medical diagnostics, the primary goal is the production of useful images in a relatively short time; modifying algorithms to accelerate convergence then becomes important. When the problem involves large-scale systems of linear equations, block-iterative methods that employ only some of the equations at each step often perform as well as simultaneous methods that use all the equations at each step, in a fraction of the time.

I have chosen to organize the topics from the general to the more specific. In recent years I have developed the class of iterative algorithms that I call the SUMMA class. These are related to sequential unconstrained minimization methods and, somewhat surprisingly, can be shown to include a wide variety of iterative algorithms well known to researchers in different fields. By unifying a variety of seemingly disparate algorithms, analogies can be considered and new properties of algorithms can be derived by analogy with known properties of other algorithms. The unification also serves to draw the attention of researchers working in one field to related algorithms in other fields, such as statisticians working on parameter estimation; image scientists processing scanning data, and mathematicians involved in theoretical and applied optimization.

Chapter 2 gives an overview of sequential optimization and the subclasses of auxiliary-function methods and the SUMMA algorithms. The next three chapters deal in greater detail with particular examples: barrier- and penalty-function methods in Chapter 3, proximal minimization in Chapter 4, and forward-backward splitting in Chapter 5. Chapter 6 through Chapter 9 focus on fixed-point algorithms for operators on Euclidean space. After that, the discussion is broadened to include distance measures other than the usual Euclidean distance. The final few chapters present specific problems to illustrate the use of iterative methods discussed previously.

The book brings together, in one place, a number of important iterative algorithms in medical imaging, optimization and statistical estimation. It includes a good deal of recent work that has not appeared in books previously. It provides a broad theoretical unification of many of these algorithms in terms of auxiliary-function methods and, in particular, the recently developed class of SUMMA algorithms. The book is somewhat limited in scope, rather than encyclopedic; the topics discussed are ones I have been personally involved with over the past couple of decades. The treatment of each topic is sufficiently detailed, without being exhaustive. Most chapters contain exercises that introduce new ideas and contribute to making the book appropriate for self study.

This book is not intended as an introduction to optimization or convex analysis, for which there are numerous texts available, such as Kelley's book [150]. Several of the topics discussed here are also treated in the books by Censor and Zenios [95], Bauschke and Combettes [18], Saad [194] and Cegielski [77]. Most of my articles in the bibliography can be downloaded from my website, http://faculty.uml.edu/cbyrne/cbyrne.html.

Chapter 1

Background

1.1 Overview

A fundamental inverse problem is the reconstruction of a function from finitely many measurements pertaining to that function. This problem is central to radar, sonar, optical imaging, transmission and emission tomography, magnetic resonance imaging, and many other applications. Because the measured data is limited, it cannot serve to determine one single correct answer. In each of these applications some sort of prior information is incorporated in the reconstruction process in order to produce a usable solution. Minimizing a cost function is a standard technique used to single out one solution from the many possibilities. The reconstruction algorithms often employ projection techniques to guarantee that the reconstructed function is consistent with the known constraints. Typical image reconstruction problems involve thousands of data values and iterative algorithms are required to perform the desired optimization.

1.1.1 Fourier-Transform Data

We begin with an example of a common remote-sensing problem in which the available data are values of the Fourier transform of the function we wish to reconstruct. In our example the function we wish to reconstruct is the amplitude function associated with a spatially extended object transmitting or reflecting electromagnetic radiation. Problems of this sort arise in a variety of applications, from mapping the sources of sunspot activity to synthetic-aperture radar and magnetic-resonance imaging. Our example is a somewhat simplified version of what is encountered in the real world, but it serves to illustrate several key aspects of most remote-sensing problems. From this example we see why it is that the data is limited, apart, of course, from the obvious need to limit ourselves to finitely many data values, and come to understand how resolution depends on the relationship between the size of the object being imaged and the frequency of the probing or transmitted signal.

Because our data is limited and the reconstruction problems are underdetermined, we are led to consider constrained optimization methods, such as constraint-consistent minimum-norm reconstructions. Once we have settled on an appropriate ambient space, usually a Hilbert space, in which to place the function to be reconstructed, it is reasonable to take as the reconstruction the data-consistent member of the space having the smallest norm. If we have additional constraints that we wish to impose, we can use orthogonal projection onto convex sets to satisfy the constraints. A key step, and one that is too often overlooked, is the choice of the ambient space. As we shall see, soft constraints coming from prior information, such as knowledge of the overall shape of the function being reconstructed, or of some prominent features of that function, can often be incorporated in the reconstruction process through the choice of the ambient space. Although Hilbert space norms are the most convenient, other Banach space norms, or distance measures not derived from norms, such as cross-entropy, can also be helpful.

It is usually the case that the function we wish to reconstruct is a real- or complex-valued function of one or more continuous variables. At some stage of the reconstruction, we must discretize the function or its estimate, if only to plot the estimate at the final step. It can be helpful to introduce the discretization earlier in the process, and most of our discussion in this book will focus on reconstructing a finite vector in $\mathbb{R}^J$ or $\mathbb{C}^J$. Once we have decided to base the reconstruction on the minimization of some cost function, we need to find an appropriate algorithm; our focus here will be on iterative minimization algorithms.

1.1.2 Transmission Tomography

Our second example is the problem of reconstructing an image from transmission tomographic data. In transmission tomography x-rays are transmitted through the object and the initial and final intensities of the x-ray beams are measured. In the continuous-variable model, the data are taken to be line integrals of the *attenuation function* to be reconstructed. In theory, if we had available all the line integrals, Fourier transform methods would provide the solution. In practice, the attenuation function is discretized. It is still possible to mimic the continuous-variable case and obtain reconstructions using discrete Fourier transformation and filtered back-projection. Alternatively, one can relate the line-integral data to a large system of linear equations to be solved using iterative methods, such as the *algebraic reconstruction technique* (ART) and its multiplicative version, the MART [134].

1.1.3 Emission Tomography

Our third example is also taken from tomography. In emission tomography, such as positron emission tomography (PET) or single-photon emission computed tomography (SPECT), a radionuclide is introduced into the body of the living object, a human being or animal, the data are counts of photons detected at gamma cameras positioned close to the body, and the reconstructed image provides an indication of where the metabolic processes have deposited the radionuclide. In the discrete model, each pixel or voxel within the body is associated with an unknown nonnegative quantity, the amount of radionuclide present at that location, which is assumed to be proportional to the expected number of photons emitted at that site during the scanning time. The randomness involved here suggests the use of statistical methods for parameter estimation, and approaches such as iterative likelihood maximization have been used for the reconstruction [193]. The *expectation maximization maximum likelihood* (EMML) algorithm converges to a maximizer of likelihood for the model of independent Poisson-distributed emitters [196, 155, 206, 156, 47].

1.2 An Urns Model for Remote Sensing

Many of the examples mentioned in the previous section are commonly described as problems of "remote sensing". In such problems what we are able to measure is not what we really want, but is related in some way to what we really want. To get a feel for remote-sensing problems without the mathematical formulation of a particular application, we consider an urns model involving urns or bowls of colored marbles. Urns models are often used to illustrate aspects of thermodynamics. Our model may seem overly simple, but, as we shall see, it accurately describes the problem we face in single-photon and positron emission tomography, SPECT and PET, respectively.

Suppose that there are J large bowls, numbered $j = 1$ through $j = J$. Each bowl contains colored marbles. Marbles come in colors with numerical codes $i = 1$ through $i = I$. For each j and i, I know the proportion of the marbles in the j-th bowl that are of the i-th color; denote this proportion by $a_{i,j} \geq 0$.

I am not interested in marbles in bowls. What I am interested in is the contents of a certain box. This box contains many small pieces of paper. On each piece of paper is written one of the j values. I want to know the proportions $x_j \geq 0$ of pieces of paper in the box that have the number j written on them. However, I have no direct access to the box.

I have an assistant, whose task it is to draw one piece of paper from the box, read the number written on it, and, out of my sight and without saying anything, go to the designated bowl, remove one marble, and announce to me its color. I write down the color. This ritual is repeated many times, after which I have a long list of colors, from which I must estimate the x_j.

Our urns model can be modified to illustrate problems dealt with by hidden Markov models [201].

1.3 Hidden Markov Models

In the urns model we just discussed, the order of the colors in the list is unimportant; we could randomly rearrange the colors on the list without affecting the nature of the problem. The probability that a green marble will be chosen next is the same, whether a blue or a red marble was just chosen the previous time. This independence from one selection to another is fine for modeling certain physical situations, such as emission tomography. However, there are other situations in which this independence does not conform to reality.

In written English, for example, knowing the current letter helps us, sometimes more, sometimes less, to predict what the next letter will be. We know that if the current letter is a "q", then there is a high probability that the next one will be a "u". So what the current letter is affects the probabilities associated with the selection of the next one.

Spoken English is even tougher. There are many examples in which the pronunciation of a certain sound is affected, not only by the sound or sounds that preceded it, but by the sound or sounds that will follow. For example, the sound of the "e" in the word "bellow" is different from the sound of the "e" in the word "below"; the sound changes, depending on whether there is a double "l" or a single "l" following the "e". Here the entire context of the letter affects its sound.

Hidden Markov models (HMM) are increasingly important in speech processing, optical character recognition and DNA sequence analysis. They allow us to incorporate dependence on the past into our model. In this section we illustrate HMM using a modification of the urns model.

Suppose, once again, that we have J urns, indexed by $j = 1, ..., J$ and I colors of marbles, indexed by $i = 1, ..., I$. In addition to the original box full of pieces of paper, now each of the J urns has its own box, and each of these boxes contains a large number of pieces of paper, with the number of one urn written on each piece. My assistant starts by drawing one piece of paper from the original box. He reads the number written on it, call it j_1, goes to the urn with the number j_1 and draws out a marble. He then announces the color. He then draws a piece of paper from box number j_1, reads the next number, say j_2, proceeds to urn number j_2, etc. After N marbles have been drawn, the only data I have is a list of colors, $\mathbf{i} = \{i_1, i_2, ..., i_N\}$.

The *transition probability* that my assistant will proceed from the urn numbered k to the urn numbered j is $b_{j,k}$, with $\sum_{j=1}^{J} b_{j,k} = 1$. The number of the current urn is the current *state*. In an ordinary *Markov chain* model, we observe directly a sequence of states governed by the transition probabilities. The Markov chain model provides a simple formalism for describing a system that moves from one state into another, as time goes on. In the hidden Markov model we are not able to observe the states directly; they are hidden from us. Instead, we have indirect observations, such as the colors of the marbles in our urns model.

The probability that the color numbered i will be drawn from the urn numbered j is $a_{i,j}$, with $\sum_{i=1}^{I} a_{,ij} = 1$, for all j. The colors announced are the *visible states*, while the unannounced urn numbers are the *hidden states.*

There are several distinct objectives one can have, when using HMM. We assume that the data is the list of colors, $\mathbf{i}$.

- **Evaluation:** For given probabilities $a_{i,j}$ and $b_{j,k}$, what is the probability that the list $\mathbf{i}$ was generated according to the HMM? Here, the objective is to see if the model is a good description of the data.
- **Decoding:** Given the model, the probabilities and the list $\mathbf{i}$, what list $\mathbf{j} = \{j_1, j_2, ..., j_N\}$ of urns is most likely to be the list of urns actually visited? Now, we want to infer the hidden states from the visible ones.
- **Learning:** We are told that there are J urns and I colors, but are not told the probabilities $a_{i,j}$ and $b_{j,k}$. We are given several data vectors $\mathbf{i}$ generated by the HMM; these are the *training sets.* The objective is to learn the probabilities.

Once again, the ML approach can play a role in solving these problems [112]. The *Viterbi algorithm* is an important tool used for the decoding phase (see [201]).

1.4 Measuring the Fourier Transform

Let $f(x) : [-L, L] \to \mathbb{C}$ have Fourier series representation

$$f(x) = \sum_{n=-\infty}^{\infty} c_n e^{in\pi x/L}, \tag{1.1}$$

where the Fourier coefficient c_n is given by

$$c_n = \frac{1}{2L} \int_{-L}^{L} f(x) e^{-in\pi x/L} dx. \tag{1.2}$$

We shall see how Fourier coefficients can arise as data obtained through measurements. However, we shall be able to measure only a finite number of the Fourier coefficients. One issue that will concern us is the effect on the approximation of $f(x)$ if we use some, but not all, of its Fourier coefficients.

1.4.1 The Discrete Fourier Transform

Suppose that we have c_n for $|n| \leq N$. It is not unreasonable to try to estimate the function $f(x)$ using the *discrete Fourier transform* (DFT) estimate, which is

$$f_{DFT}(x) = \sum_{n=-N}^{N} c_n e^{in\pi x/L}. \tag{1.3}$$

In Figure 1.1 below, the function $f(x)$ is the solid-line figure in both graphs. In the bottom graph, we see the true $f(x)$ and a DFT estimate. The top graph is the result of *band-limited extrapolation*, a technique for predicting missing Fourier coefficients that we shall discuss later.

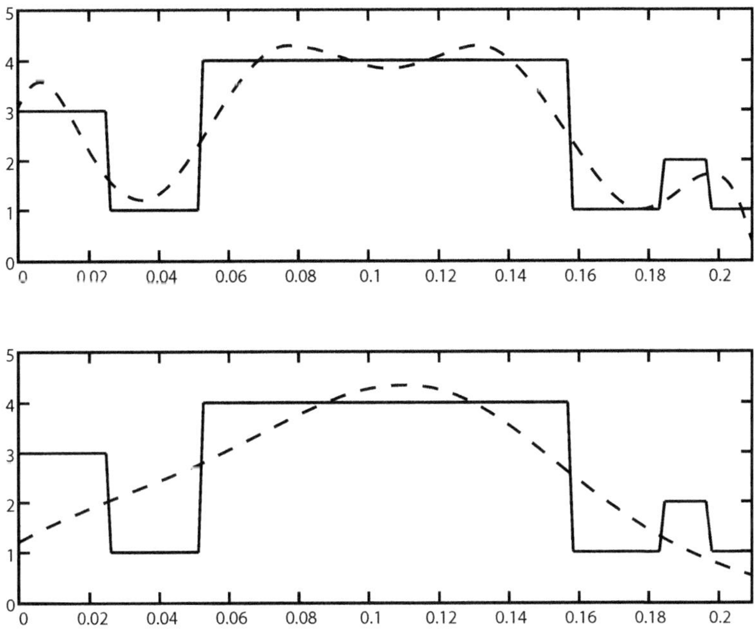

FIGURE 1.1: The non-iterative band-limited extrapolation method (MDFT) (top) and the DFT (bottom) for 30 times over sampled data. The solid line is the true object.

1.4.2 The Unknown Amplitude Problem

In this example, we imagine that each point x in the interval $[-L, L]$ is sending a signal at the frequency ω, each with its own amplitude $f(x)$;

that is, the signal sent by the point x is

$$f(x)e^{i\omega t}; \tag{1.4}$$

here the amplitude contains both magnitude and phase, so is complex. We imagine that the amplitude function $f(x)$ is unknown and we want to determine it. It could be the case that the signals originate at the points x, as with light or radio waves from the sun, or are simply reflected from the points x, as is sunlight from the moon or radio waves in radar.

Now let us consider what is received by a point P on the circumference of a circle centered at the origin and having large radius D. The point P corresponds to the angle θ as shown in Figure 1.2. It takes a finite time for the signal sent from x at time t to reach P, so there is a delay.

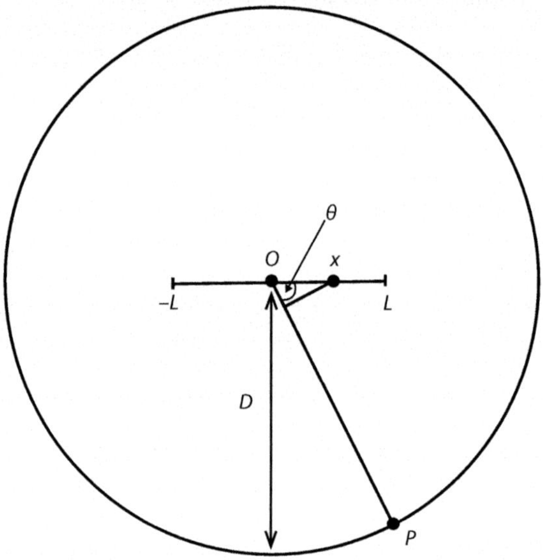

FIGURE 1.2: Far-field Measurements. The distance from x to P is approximately $D - x\cos\theta$.

We assume that c is the speed at which the signal propagates. Because D is large relative to L, we make the *far-field assumption*, which allows us to approximate the distance from x to P by $D - x\cos(\theta)$. Therefore, what P receives at time t from x is approximately what was sent from x at time $t - \frac{1}{c}(D - x\cos(\theta))$.

At time t, the point P receives from x the signal

$$f(x)e^{i\omega(t-\frac{1}{c}(D-x\cos(\theta)))}, \tag{1.5}$$

or

$$e^{i\omega(t-\frac{1}{c}D)}f(x)e^{i\omega x\cos(\theta)/c}. \tag{1.6}$$

Therefore, from our measurement at P, we obtain

$$e^{i\omega(t-\frac{1}{c}D)}\int_{-L}^{L} f(x)e^{i\omega x\cos(\theta)/c}dx. \tag{1.7}$$

Consequently, from measurements in the far field we obtain the values

$$\int_{-L}^{L} f(x)e^{i\omega x\cos(\theta)/c}dx, \tag{1.8}$$

where θ can be chosen as any angle between 0 and 2π. When we select θ so that

$$\frac{\omega\cos(\theta)}{c} = \frac{n\pi}{L}, \tag{1.9}$$

we have c_{-n}.

1.4.3 Limited Data

Note that we will be able to solve Equation (1.9) for θ only if we have

$$|n| \leq \frac{L\omega}{\pi c}. \tag{1.10}$$

This tells us that we can measure only finitely many of the Fourier coefficients of $f(x)$. It is common in signal processing to speak of the *wavelength* of a sinusoidal signal; the wavelength associated with a given ω and c is

$$\lambda = \frac{2\pi c}{\omega}. \tag{1.11}$$

Therefore, we can measure c_n for $|n|$ not greater than $\frac{2L}{\lambda}$, which is the length of the interval $[-L, L]$, measured in units of wavelength λ. We get more Fourier coefficients when the product $L\omega$ is larger; this means that when L is small, we want ω to be large, so that λ is small and we can measure more Fourier coefficients. As we saw previously, using these finitely many Fourier coefficients to calculate the DFT reconstruction of $f(x)$ can lead to a poor estimate of $f(x)$, particularly when we don't have many Fourier coefficients.

1.4.4 Can We Get More Information?

As we just saw, we can make measurements at any point P in the far field; perhaps we do not need to limit ourselves to just those angles that lead to the limited number of Fourier coefficients c_n.

We define the Fourier transform of the function $f(x)$ to be the function

$$F(\gamma) = \int_{-L}^{L} f(x)e^{i\gamma x}dx. \tag{1.12}$$

Therefore, when we measure the signals received at the point P in the far field, we obtain the value $F(\gamma)$ for $\gamma = \omega\cos(\theta)/c$. Therefore, in principle, we have available to us all the values of $F(\gamma)$ for γ in the interval $[-\omega/c, \omega/c]$. These are not all of the non-zero values of $F(\gamma)$, of course, since $F(\gamma)$ is band-limited, but not support-limited.

1.4.5 Over-Sampling

It is sometimes argued that once we have obtained all the values of c_n that are available to us, there is no more information about $f(x)$ that we can obtain through further measurements in the far field; this is wrong. It may come as somewhat of a surprise, but from the theory of complex analytic functions, or from band-limited extrapolation [71, 72], we can prove that there is enough data available to us here to reconstruct $f(x)$ perfectly, at least in principle. The drawback, in practice, is that the measurements would have to be free of noise and impossibly accurate. All is not lost, however.

Suppose, for the sake of illustration, that we measure the far-field signals at points P corresponding to angles θ that satisfy

$$\frac{\omega\cos(\theta)}{c} = \frac{n\pi}{2L}, \tag{1.13}$$

instead of

$$\frac{\omega\cos(\theta)}{c} = \frac{n\pi}{L}.$$

Now we have twice as many data points and from our new measurements we can obtain

$$a_m = \frac{1}{4L}\int_{-L}^{L} f(x)e^{-ix\frac{m\pi}{2L}}dx = \frac{1}{4L}\int_{-2L}^{2L} f(x)e^{-ix\frac{m\pi}{2L}}dx, \tag{1.14}$$

for $|m| \leq M$, which are Fourier coeffcients of $f(x)$ when viewed as a function defined on the interval $[-2L, 2L]$, but still zero outside $[-L, L]$. We say now that our data is *twice over-sampled.* Note that we call it *over-sampled* because the rate at which we are sampling is higher, even though the distance between samples is lower.

For clarity, let us denote the function defined on the interval $[-2L, 2L]$ that equals $f(x)$ for x in $[-L, L]$ and is equal to zero elsewhere as $g(x)$. We have twice the number of Fourier coefficients that we had previously, but for the function $g(x)$. A DFT reconstruction using this larger set of

Fourier coefficients will reconstruct $g(x)$ on the interval $[-2L, 2L]$; this DFT estimate is

$$g_{DFT}(x) = \sum_{m=-M}^{M} a_m e^{im\pi x/2L}, \tag{1.15}$$

for $|x| \leq 2L$. This will give us a reconstruction of $f(x)$ itself over the interval $[-L, L]$, but will also give us a reconstruction of the rest of $g(x)$, which we already know to be zero. So we are wasting the additional data by reconstructing $g(x)$ instead of $f(x)$. We need to use our prior knowledge that $g(x) = 0$ for $L < |x| \leq 2L$.

We want to use the prior knowledge that $g(x) = 0$ for $L < |x| \leq 2L$ to improve our reconstruction. Suppose that we take as our reconstruction the *modified DFT* (MDFT) [40]:

$$f_{MDFT}(x) = \sum_{j=-M}^{M} b_j e^{ij\pi x/2L}, \tag{1.16}$$

for $|x| \leq L$, and zero elsewhere, with the b_j chosen so that $f_{MDFT}(x)$ is consistent with the measured data. Calculating this estimator involves solving a system of linear equations for the b_j.

The top graph in Figure 1.1 illustrates the improvement over the DFT that can be had using the MDFT. In that figure, we took data that was thirty times over-sampled, not just twice over-sampled, as in our previous discussion. Consequently, we had thirty times the number of Fourier coefficients we would have had otherwise, but for an interval thirty times longer. To get the top graph, we used the MDFT, with the prior knowledge that $f(x)$ was non-zero only within the central thirtieth of the long interval. The bottom graph shows the DFT reconstruction using the larger data set, but only for the central thirtieth of the full period, which is where the original $f(x)$ is non-zero.

1.4.6 A Projection-Based View

When we view the function $f(x)$ as a member of the Hilbert space $L^2(-L, L)$, we find that the DFT estimate of $f(x)$ is the orthogonal projection of the zero function onto the closed convex subset of all members of $L^2(-L, L)$ that are consistent with the data; that is, the DFT estimate is the member of $L^2(-L, L)$ that has minimum norm among all those members consistent with the data. The MDFT estimate is the member of $L^2(-2L, 2L)$ of minimum norm among all members that are both consistent with the data and supported on the interval $[-L, L]$. The MDFT estimate is also the member of $L^2(-L, L)$ of minimum norm consistent with the over-sampled data. The MDFT is not the DFT in this case, since the functions $e^{ij\pi x/2L}$ are not orthogonal with respect to the usual inner product on $L^2(-L, L)$.

1.4.7 Other Forms of Prior Knowledge

As we just showed, knowing that we have over-sampled in our measurements can help us improve the resolution in our estimate of $f(x)$. We may have other forms of prior knowledge about $f(x)$ that we can use. If we know something about large-scale features of $f(x)$, but not about finer details, we can use the PDFT estimate, which is a generalization of the MDFT [41, 42].

We can write the MDFT estimate above as

$$f_{MDFT}(x) = \chi_{[-L,L]}(x) \sum_{j=-M}^{M} b_j e^{ij\pi x/2L}; \tag{1.17}$$

here $\chi_{[-L,L]}(x)$ is one for $|x| \leq L$, and zero, otherwise. Written this way, we see that the second factor has the algebraic form of the DFT estimate, while the first factor incorporates our prior knowledge that $f(x)$ is zero for $|x| > L$.

Suppose that we have some prior knowledge of the function $|f(x)|$ beyond simply support information. Let us select $p(x) > 0$ as a prior estimate of $|f(x)|$ and let our PDFT estimate of $f(x)$ have the form

$$f_{PDFT}(x) = p(x) \sum_{j=-M}^{M} d_j e^{ij\pi x/2L}, \tag{1.18}$$

with the coefficients d_j computed by forcing $f_{PDFT}(x)$ to be consistent with the measured data. Again, this involves solving a system of linear equations, although there are other ways to handle this. By discretizing the problem, the PDFT can be calculated using the ART algorithm discussed below [197, 198]. The PDFT approach extends to higher dimensions, as we illustrate in the following example.

The original image on the upper right of Figure 1.3 is a discrete rectangular array of intensity values simulating a slice of a head. The data was obtained by taking the two-dimensional discrete Fourier transform of the original image, and then discarding, that is, setting to zero, all these spatial frequency values, except for those in a smaller rectangular region around the origin. The problem then is under-determined. A minimum-norm solution would seem to be a reasonable reconstruction method.

The DFT reconstruction is the minimum-two-norm solution shown on the lower right. It is calculated simply by performing an inverse discrete Fourier transform on the array of retained discrete Fourier transform values. The original image has relatively large values where the skull is located,

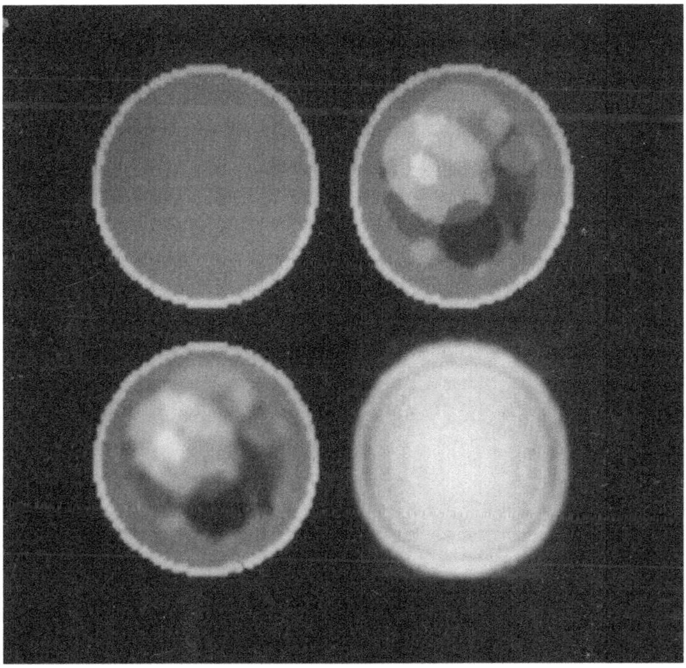

FIGURE 1.3: Extracting information in image reconstruction The original is on the upper right. The lower right is the minimum-two-norm (DFT) solution, and the lower left is a minimum weighted-two-norm (PDFT) solution. The prior estimate is on the upper left.

but the minimum-norm reconstruction does not want such high values; the norm involves the sum of squares of intensities, and high values contribute disproportionately to the norm. Consequently, the minimum-norm reconstruction chooses instead to conform to the measured data by spreading what should be the skull intensities throughout the interior of the skull. The minimum-norm reconstruction does tell us something about the original; it tells us about the existence of the skull itself, which, of course, is indeed a prominent feature of the original. However, in all likelihood, we would already know about the skull; it would be the interior that we want to know about.

Using our knowledge of the presence of a skull, which we might have obtained from the minimum-norm reconstruction itself, we construct the prior estimate shown in the upper left. Now we use the same data as before, and calculate a minimum-weighted-norm (PDFT) reconstruction, using as the weight vector the reciprocals of the values of the prior image. This minimum-weighted-norm reconstruction is shown on the lower left;

it is clearly almost the same as the original image. The calculation of the minimum-weighted norm solution can be done iteratively using the ART algorithm [198].

When we weight the skull area with the inverse of the prior image, we allow the reconstruction to place higher values there without having much of an effect on the overall weighted norm. In addition, the reciprocal weighting in the interior makes spreading intensity into that region costly, so the interior remains relatively clear, allowing us to see what is really present there.

When we try to reconstruct an image from limited data, it is easy to assume that the information we seek has been lost, particularly when a reasonable reconstruction method fails to reveal what we want to know. As this example, and many others, show, the information we seek is often still in the data, but needs to be brought out in a more subtle way.

1.5 Transmission Tomography

The ART and the MART are two iterative algorithms that were designed to address issues that arose in solving large-scale systems of linear equations for medical imaging [134]. The EMART is a more recently discovered method that combines useful features of both ART and MART [50]. In this chapter we give an overview of ART and MART; we shall revisit them later in more detail.

1.5.1 The ART and MART

In many applications, such as in image processing, we need to solve a system of linear equations that is quite large, often several tens of thousands of equations in about the same number of unknowns. In these cases, issues such as the costs of storage and retrieval of matrix entries, the computation involved in apparently trivial operations, such as matrix-vector products, and the speed of convergence of iterative methods demand greater attention. At the same time, the systems to be solved are often underdetermined, and solutions satisfying certain additional constraints, such as nonnegativity, are required.

Both the *algebraic reconstruction technique* (ART) and the *multiplicative algebraic reconstruction technique* (MART) were introduced as two iterative methods for discrete image reconstruction in transmission tomography.

Both methods are what are called *row-action* methods, meaning that each step of the iteration uses only a single equation from the system. The MART is limited to nonnegative systems for which nonnegative solutions are sought. In the under-determined case, both algorithms find the solution closest to the starting vector, in the two-norm or weighted two-norm sense for ART, and in the cross-entropy sense for MART, so both algorithms can be viewed as solving optimization problems. In Chapter 14 we describe the use of MART to solve the dual geometric programming problem. For both algorithms, the starting vector can be chosen to incorporate prior information about the desired solution. In addition, the ART can be employed in several ways to obtain a least-squares solution, in the over-determined case.

The *simultaneous* MART (SMART) algorithm is a simultaneous variant of the MART in which all the equations are employed at each step of the iteration. Closely related to the SMART is the *expectation maximization maximum likelihood* (EMML) method, which is also a simultaneous algorithm.

The EMART is a row-action variant of the EMML algorithm. Like MART, it applies to nonnegative systems of equations and produces nonnegative solutions, but, like ART, does not require exponentiation, so is computationally simpler than MART.

1.5.2 The ART in Tomography

In x-ray transmission tomography, as an x-ray beam passes through the body, it encounters various types of matter, such as soft tissue, bone, ligaments, air, each weakening the beam to a greater or lesser extent. If the intensity of the beam upon entry is I_{in} and I_{out} is its lower intensity after passing through the body, then, at least approximately,

$$I_{out} = I_{in} e^{-\int_L f},$$

where $f = f(x, y) \geq 0$ is the *attenuation function* describing the two-dimensional distribution of matter within the slice of the body being scanned and $\int_L f$ is the integral of the function f over the line L along which the x-ray beam has passed. This is the continuous model. In the discrete model the slice of the body being scanned is viewed as consisting of pixels, which we number $j = 1, 2, ..., J$. The x-rays are sent into the body along I lines, which we number $i = 1, 2, ..., I$. The line integral of f along the i-th line is measured, approximately, from the entering and exiting strengths of the x-ray beams; these measurements are denoted b_i.

For $i = 1, ..., I$, let L_i be the set of pixel indices j for which the j-th pixel intersects the i-th line segment, as shown in Figure 1.4, and let $|L_i|$ be the cardinality of the set L_i. Let $A_{ij} = 1$ for j in L_i, and $A_{ij} = 0$ otherwise.

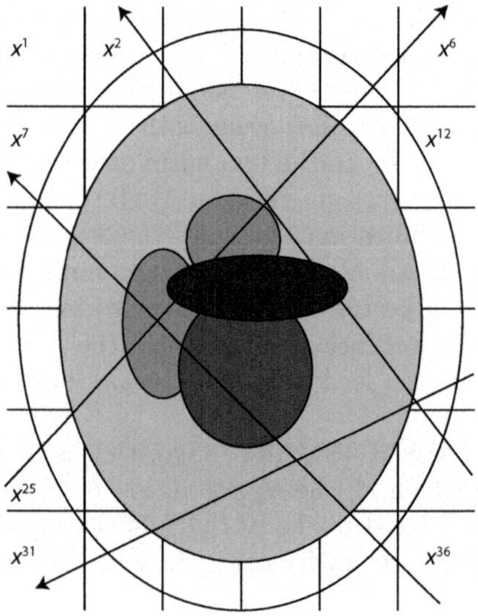

FIGURE 1.4: Line integrals through a discretized object.

With $i = k(\text{mod}\, I) + 1$, the iterative step of the ART algorithm is

$$x_j^{k+1} = x_j^k + \frac{1}{|L_i|}(b_i - (Ax^k)_i), \tag{1.19}$$

for j in L_i, and

$$x_j^{k+1} = x_j^k, \tag{1.20}$$

if j is not in L_i. In each step of ART, we take the error, $b_i - (Ax^k)_i$, associated with the current x^k and the i-th equation, and distribute it equally over each of the pixels that intersects L_i.

This model is too simple; we are assuming that if the line segment intersects a pixel, then the entire amount of attenuating material within that pixel affects the x-ray strength. A somewhat more sophisticated version of ART allows A_{ij} to include the length of the i-th line segment that lies within the j-th pixel; A_{ij} is taken to be the ratio of this length to the length of the diagonal of the j-th pixel.

More generally, ART can be viewed as an iterative method for solving an arbitrary system of linear equations, $Ax = b$.

1.5.3 The ART in the General Case

Throughout this book we denote by $\mathbb{R}^J$ and $\mathbb{C}^J$ the vectors spaces of J-dimensional column vectors whose entries are real or complex numbers, respectively. For any vectors x and y in $\mathbb{C}^J$ the inner product is

$$\langle x, y \rangle = x \cdot y = y^\dagger x = \sum_{j=1}^{J} x_j \overline{y_j}, \tag{1.21}$$

where $y^\dagger$ is the conjugate transpose of the column vector y. The Euclidean norm or two-norm of x is

$$\|x\|_2 = \sqrt{\langle x, x \rangle}, \tag{1.22}$$

and the Euclidean distance between x and y is $\|x - y\|_2$.

The Euclidean norm is not the only norm on $\mathbb{C}^J$ that interests us. We shall also make use of the one-norm, given by

$$\|x\|_1 = \sum_{j=1}^{J} |x_j|. \tag{1.23}$$

Let A be a matrix with complex entries, having I rows and J columns, and let b be a member of $\mathbb{C}^I$. We want to solve the system $Ax = b$. Note that when we say that A is a complex matrix and b a complex vector, we do not exclude the case in which the entries of both A and b are real.

Associated with each equation $(Ax)_i = b_i$ in the system $Ax = b$ there is a hyperplane H_i defined to be the subset of J-dimensional column vectors given by

$$H_i = \{x | (Ax)_i = b_i\}. \tag{1.24}$$

Definition 1.1 *The* orthogonal projection operator *onto the hyperplane H_i is the function $P_i : \mathbb{C}^J \to \mathbb{C}^J$ defined for each z in $\mathbb{C}^J$ by $P_i z = x$, where x is the member of H_i closest to z, in the sense of the Euclidean distance.*

The ART algorithm can be expressed in terms of the operators P_i. Let x^0 be arbitrary and, for each nonnegative integer k, let $i(k) = k(\text{mod}\, I) + 1$. The iterative step of the ART is

$$x^{k+1} = P_{i(k)} x^k. \tag{1.25}$$

We can write the iterative step of the ART explicitly, as follows.

Algorithm 1.1 (ART) *Let $\alpha_i = \sum_{j=1}^{J} |A_{ij}|^2$. For $k = 0, 1, ...$ and $i = i(k) = k(\text{mod}\, I) + 1$, the entries of x^{k+1} are*

$$x_j^{k+1} = x_j^k + \alpha_i^{-1} \overline{A_{ij}} (b_i - (Ax^k)_i). \tag{1.26}$$

Because the ART uses only a single equation at each step, it has been called a *row-action* [80] or *sequential* method.

1.5.3.1 When $Ax = b$ Has Solutions

For the consistent case we have the following result concerning the ART.

Theorem 1.1 *Let $A\hat{x} = b$ and let x^0 be arbitrary. Let $\{x^k\}$ be generated by the ART. Then the sequence of Euclidean distances $\{||\hat{x} - x^k||_2\}$ is decreasing and $\{x^k\}$ converges to the solution of $Ax = b$ closest to x^0.*

This theorem is a particular case of the more general Theorem 8.3.

So, when the system $Ax = b$ has exact solutions, the ART converges to the solution closest to x^0, in the Euclidean distance. How fast the algorithm converges will depend on the ordering of the equations and on whether or not we use *relaxation*, which we shall discuss later. In selecting the equation ordering, the important thing is to avoid particularly bad orderings, in which the hyperplanes H_i and H_{i+1} are nearly parallel [141].

1.5.3.2 When $Ax = b$ Has No Solutions

When there are no exact solutions, the ART does not converge to a single vector, but, for each fixed i, the subsequence $\{x^{nI+i}, n = 0, 1, ...\}$ converges to a vector z^i and the collection $\{z^i\,|i = 1, ..., I\}$ is called the *limit cycle* [202]. The ART limit cycle will vary with the ordering of the equations, and contains more than one vector unless an exact solution exists. Figures 1.5 and 1.6 illustrate the behavior of the ART in the two cases.

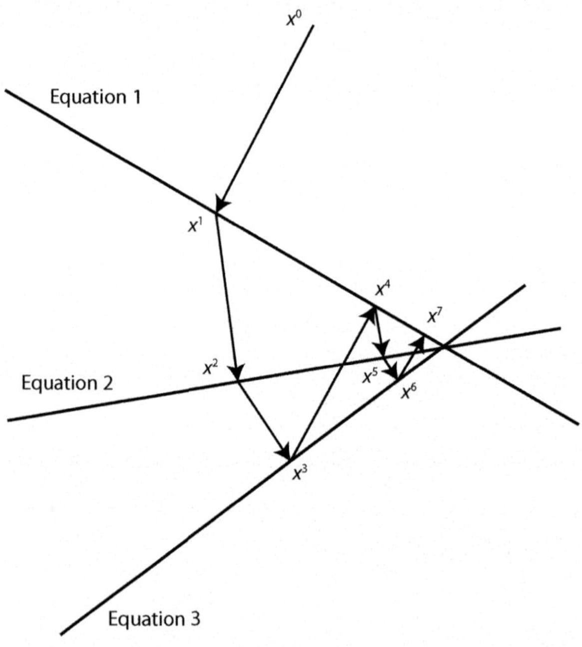

FIGURE 1.5: The ART algorithm in the consistent case.

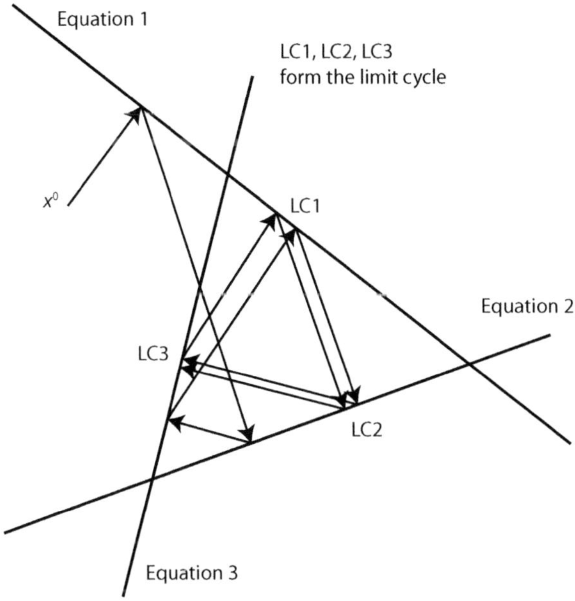

FIGURE 1.6: The ART algorithm in the inconsistent case, illustrating subsequential convergence to a limit cycle.

1.5.4 The MART

The *multiplicative* ART (MART) is an iterative algorithm closely related to the ART. It also was devised to obtain tomographic images, but, like ART, applies more generally; MART applies to nonnegative systems of linear equations $Ax = b$ for which the b_i are positive, the A_{ij} are nonnegative, and the solution x we seek is to have nonnegative entries. It is not so easy to see the relation between ART and MART if we look at the most general formulation of MART. For that reason, we begin with a simpler case, transmission tomographic imaging, in which the relation is most clearly apparent.

1.5.4.1 A Special Case of MART

We begin by considering the application of MART to the transmission tomography problem. Once again, for $i = 1, ..., I$, let L_i be the set of pixel indices j for which the j-th pixel intersects the i-th line segment, and let $|L_i|$ be the cardinality of the set L_i. Let $A_{ij} = 1$ for j in L_i, and $A_{ij} = 0$ otherwise. In each step of ART, we take the error, $b_i - (Ax^k)_i$, associated with the current x^k and the i-th equation, and distribute it equally over each of the pixels that intersects L_i. Suppose, now, that each b_i is positive, and we know in advance that the desired image we wish to reconstruct must be nonnegative. We can begin with $x^0 > 0$, but as we compute the ART steps, we may lose nonnegativity. One way to avoid this loss is to correct the current x^k multiplicatively, rather than additively, as in ART. This leads to the *multiplicative* ART (MART).

The MART, in this case, has the iterative step

$$x_j^{k+1} = x_j^k \Big(\frac{b_i}{(Ax^k)_i} \Big), \tag{1.27}$$

for those j in L_i, and

$$x_j^{k+1} = x_j^k, \tag{1.28}$$

otherwise. Therefore, we can write the iterative step as

$$x_j^{k+1} = x_j^k \Big(\frac{b_i}{(Ax^k)_i} \Big)^{A_{ij}}. \tag{1.29}$$

1.5.4.2 The MART in the General Case

Taking the entries of the matrix A to be either one or zero, depending on whether or not the j-th pixel is in the set L_i, is too crude. The line L_i may just clip a corner of one pixel, but pass through the center of another. Surely, it makes more sense to let A_{ij} be the length of the intersection of line L_i with the j-th pixel, or, perhaps, this length divided by the length of the diagonal of the pixel. It may also be more realistic to consider a strip, instead of a line. Other modifications to A_{ij} may be made, in order to better describe the physics of the situation. Finally, all we can be sure of is that A_{ij} will be nonnegative, for each i and j. In such cases, what is the proper form for the MART?

The MART, which can be applied only to nonnegative systems, is a sequential, or row-action, method that uses one equation only at each step of the iteration.

Algorithm 1.2 (MART) *Let x^0 be a positive vector. For $k = 0, 1, ...,$ and $i = k(\operatorname{mod} I) + 1$, having found x^k define x^{k+1} by*

$$x_j^{k+1} = x_j^k \Big(\frac{b_i}{(Ax^k)_i} \Big)^{m_i^{-1} A_{ij}}, \tag{1.30}$$

where $m_i = \max\{A_{ij} \,|j = 1, 2, ..., J\}$.

Some treatments of MART leave out the m_i, but require only that the entries of A have been rescaled so that $A_{ij} \leq 1$ for all i and j. The m_i is important, however, in accelerating the convergence of MART.

Notice that we can write x_j^{k+1} as a weighted geometric mean of x_j^k and $x_j^k\left(\frac{b_i}{(Ax^k)_i}\right)$:

$$x_j^{k+1} = \left(x_j^k\right)^{1-m_i^{-1}A_{ij}} \left(x_j^k\left(\frac{b_i}{(Ax^k)_i}\right)\right)^{m_i^{-1}A_{ij}}. \tag{1.31}$$

This will help to motivate the EMART.

1.5.4.3 Cross-Entropy

For $a > 0$ and $b > 0$, let the cross-entropy or Kullback-Leibler (KL) distance [153] from a to b be

$$KL(a, b) = a \log \frac{a}{b} + b - a, \tag{1.32}$$

with $KL(a, 0) = +\infty$, and $KL(0, b) = b$. Extend to nonnegative vectors coordinate-wise, so that

$$KL(x, z) = \sum_{j=1}^{J} KL(x_j, z_j). \tag{1.33}$$

Then $KL(x, z) \geq 0$ and $KL(x, z) = 0$ if and only if $x = z$.

Unlike the Euclidean distance, the KL distance is not symmetric; $KL(Ax, b)$ and $KL(b, Ax)$ are distinct, and we can obtain different approximate solutions of $Ax = b$ by minimizing these two distances with respect to nonnegative x. We discuss this point further in Chapter 11.

1.5.4.4 Convergence of MART

In the consistent case, by which we mean that $Ax = b$ has nonnegative solutions, we have the following convergence theorem for MART.

Theorem 1.2 *In the consistent case, the MART converges to the unique nonnegative solution of $b = Ax$ for which the distance $KL(x, x^0)$ is minimized.*

If the starting vector x^0 is the vector whose entries are all one, then the MART converges to the solution that maximizes the *Shannon entropy*,

$$SE(x) = \sum_{j=1}^{J} x_j \log x_j - x_j. \tag{1.34}$$

This theorem is a particular case of more general results established in Chapter 11. As with ART, the speed of convergence is greatly affected by the ordering of the equations, converging most slowly when consecutive equations correspond to nearly parallel hyperplanes.

Open Question: When there are no nonnegative solutions, MART does not converge to a single vector, but, like ART, is always observed to produce a limit cycle of vectors. Unlike ART, there is no proof, so far, of the existence of a limit cycle for MART.

1.6 Emission Tomography

In our third example we focus on SPECT, which is somewhat simpler to describe than PET, although the reconstruction problems are essentially the same. We take $x_j \geq 0$, $j = 1, ..., J$, to be the unknown concentrations of radionuclide at the j-th pixel, and assume that x_j is also the expected number of photons emitted at the j-th pixel during the scanning time. For $i = 1, ..., I$, the random variable Y_i is the expected number of photons detected at the i-th gamma camera during the scanning, the quantity $y_i > 0$ is the actual photon count, and $P_{i,j}$ is the probability that a photon emitted from pixel j will be detected at detector i. The entries of the matrix $P = [P_{i,j}]$ are nonnegative and we assume that $s_j = \sum_{i=1}^{I} P_{i,j} > 0$, for all j. It is assumed that the random variables Y_i are independent, and each is Poisson-distributed, with mean $(Px)_i$, where $x = (x_1, ..., x_J)^T$ is the vector of unknown intensities. The entries of x are taken to be parameters to be estimated by likelihood maximization.

As promised, we relate the SPECT problem to our bowls model of remote sensing. The pixels correspond to the bowls, and the colors to the detectors. For each j and i we assume that we know, or can approximate reasonably well, the probability that a photon emitted from the j-th pixel will be detected at the i-th detector. What we want is the probability $x_j \geq 0$ that nature will select the j-th pixel to be the site of the next photon emission. This probability we assume is proportional to the relative intensity of the radionuclide within the j-th pixel, as compared with the other pixels. What we are able to measure is the number of times each detector records a photon arrival. We have a long list of detector numbers now, in place of the long list of colors.

1.6.1 The EMML Algorithm

The *expectation maximization* (EM) approach to likelihood maximization [108] is not a single algorithm, but a template for the design of algorithms. For the SPECT case, the EM approach leads to the EMML algorithm. Having selected a positive starting vector x^0, and having calculated x^k, the next iterate x^{k+1} is found using

$$x_j^{k+1} = x_j^k s_j^{-1} \sum_{i=1}^{I} P_{i,j} \frac{y_i}{(Px^k)_i}, \tag{1.35}$$

for each j. As we shall show in Chapter 11, for any $x^0 > 0$, the sequence $\{x^k\}$ converges to a maximizer of the likelihood. It is reasonable to ask if there is any connection between the ART and the EMML.

1.6.2 Relating the ART and the EMML

Both the ART and the EMML algorithm are used for medical image reconstruction, but there the resemblance seems to end. The ART is a sequential algorithm, using only one data value at a time, while the EMML is simultaneous, using all the data at each step. The EMML has its roots in statistical parameter estimation, while the ART is a deterministic method for solving systems of linear equations. The ART can be used to solve any system of linear equations, while the solutions sought using the EMML method must be nonnegative vectors. The ART employs orthogonal projection onto hyperplanes, while the EMML algorithm is best studied using the Kullback-Leibler, or cross-entropy, measure of distance. The ART converges relatively quickly, while the EMML is known to be slow.

The first step in connecting the ART and the EMML algorithm is to formulate the EMML as a method for solving a system of linear equations. The Kullback-Leibler distance is essential here. Maximizing the likelihood in the SPECT case is equivalent to minimizing $KL(y, Px)$ over all nonnegative vectors x, where $y = (y_1, ..., y_I)^T$. Therefore, the EMML algorithm can be viewed as a general iterative method for finding an exact or approximate nonnegative solution for a nonnegative system of linear equations. The ART is a sequential algorithm, but it has simultaneous versions, Cimmino's algorithm and the more general Landweber and projected Landweber methods.

The MART provides a second link between the ART and the EMML algorithm. Like the EMML, the MART can be viewed as a method for solving nonnegative systems of linear equations. Like the EMML, the properties of the MART are best revealed using the KL distance. Finally, while the MART is a sequential algorithm, it has a simultaneous version, the SMART [106, 195, 93, 47]. By developing the SMART and the EMML in tandem, as in Chapter 11 (see also [49]), we can see just how closely related these algorithms are. While the EMML minimizes $KL(y, Px)$, the SMART can be shown to minimize $KL(Px, y)$.

1.7 A Unifying Framework

The Landweber algorithm minimizes the function $f(x) = \frac{1}{2}\|Ax - b\|_2^2$, and converges to the least-squares solution of $Ax = b$ closest to the starting vector x^0 in the Euclidean distance. The EMML minimizes $f(x) = KL(y, Px)$, while the SMART minimizes $f(x) = KL(Px, y)$. All of these algorithms are sequential optimization methods, in the sense that one difficult minimization problem is replaced by a sequence of simpler ones. At each step of the iteration, we minimize a function of the form $f(x) + g_k(x)$, where the $g_k(x)$ can be chosen to permit the next iterate to be calculated in closed form, to impose constraints or penalize violations of the constraints, and to control the behavior of the sequence $\{f(x^k)\}$. In Chapter 2 we give some examples of sequential optimization, define the subclasses of auxiliary-function and SUMMA algorithms, and present brief discussions of several topics to be considered in more detail in subsequent chapters.

Chapter 2

Sequential Optimization

The Landweber and projected Landweber algorithms, the SMART and the EMML are all examples of *sequential optimization* methods. Perhaps the best known examples of sequential optimization are the *sequential unconstrained minimization* (SUM) methods [126]. Auxiliary-function algorithms, a broad subclass of sequential optimization methods, provide a unifying framework for these and many other iterative algorithms. In this chapter we consider examples of SUM methods, define the AF and SUMMA classes of algorithms, and present brief discussions of several topics to be considered in more detail in subsequent chapters.

2.1 Overview

Consider the problem of optimizing a real-valued function f over a subset C of an arbitrary set X. There may well be no simple way to solve this problem and iterative methods may be required. Many well known iterative optimization methods can be described as *sequential optimization* methods. In such methods we replace the original problem with a sequence of simpler optimization problems, obtaining a sequence $\{x^k\}$ of members of the set X. Our hope is that this sequence $\{x^k\}$ will converge to a solution of the original problem, which, of course, will require a topology on X. We may lower our expectations and ask only that the sequence $\{f(x^k)\}$ converge to $d = \inf_{x \in C} f(x)$. Failing that, we may ask only that the sequence $\{f(x^k)\}$ be nonincreasing. One way to design a sequential optimization algorithm is to use auxiliary functions. At the kth step of the iteration we minimize a function

$$G_k(x) = f(x) + g_k(x), \tag{2.1}$$

to obtain x^k.

In SUM methods the auxiliary functions $g_k(x)$ are selected to enforce the constraint that x be in C, as in barrier-function methods, or to penalize violations of that constraint, such as in penalty-function methods.

Auxiliary-function (AF) methods, which we shall discuss in some detail, closely resemble SUM methods. In AF methods certain restrictions are placed on the auxiliary functions $g_k(x)$ to control the behavior of the sequence $\{f(x^k)\}$. Even when there are no constraints, the problem of minimizing a real-valued function may require iteration; the formalism of AF minimization can be useful in deriving such iterative algorithms, as well as in proving convergence. As originally formulated, barrier- and penalty-function algorithms are not in the AF class, but can be reformulated as AF algorithms.

In AF methods the auxiliary functions satisfy additional properties that guarantee that the sequence $\{f(x^k)\}$ is nonincreasing. To have the sequence $\{f(x^k)\}$ converging to d we need to impose an additional condition on the $g_k(x)$, the SUMMA condition. The SUMMA condition may seem quite restrictive and ad hoc, and the resulting SUMMA class of algorithms fairly limited, but this is not the case. Many of the best known iterative optimization methods either are in the SUMMA class, or, like the barrier- and penalty-function methods, can be reformulated as SUMMA algorithms.

2.2 Examples of SUM

Barrier-function algorithms and penalty-function algorithms are two of the best known examples of SUM.

2.2.1 Barrier-Function Methods

Suppose that $C \subseteq \mathbb{R}^J$ and $b : C \to \mathbb{R}$ is a barrier function for C, that is, b has the property that $b(x) \to +\infty$ as x approaches the boundary of C. At the kth step of the iteration we minimize

$$B_k(x) = f(x) + \frac{1}{k}b(x) \tag{2.2}$$

to get x^k. Then each x^k is in C. We want the sequence $\{x^k\}$ to converge to some x^* in the closure of C that solves the original problem. Barrier-function methods are *interior-point methods* because each x^k satisfies the constraints.

For example, suppose that we want to minimize the function $f(x) = f(x_1, x_2) = x_1^2 + x_2^2$, subject to the constraint that $x_1 + x_2 \geq 1$. The constraint is then written $g(x_1, x_2) = 1 - (x_1 + x_2) \leq 0$. We use the logarithmic barrier function $b(x) = -\log(x_1 + x_2 - 1)$. For each positive integer k, the vector $x^k = (x_1^k, x_2^k)$ minimizing the function

$$B_k(x) = x_1^2 + x_2^2 - \frac{1}{k}\log(x_1 + x_2 - 1) = f(x) + \frac{1}{k}b(x)$$

has entries

$$x_1^k = x_2^k = \frac{1}{4} + \frac{1}{4}\sqrt{1 + \frac{4}{k}}.$$

Notice that $x_1^k + x_2^k > 1$, so each x^k satisfies the constraint. As $k \to +\infty$, x^k converges to $(\frac{1}{2}, \frac{1}{2})$, which is the solution to the original problem. The use of the logarithmic barrier function forces $x_1 + x_2 - 1$ to be positive, thereby enforcing the constraint on $x = (x_1, x_2)$.

2.2.2 Penalty-Function Methods

Again, our goal is to minimize a function $f : \mathbb{R}^J \to \mathbb{R}$, subject to the constraint that $x \in C$, where C is a nonempty closed subset of $\mathbb{R}^J$. We select a nonnegative function $p : \mathbb{R}^J \to \mathbb{R}$ with the property that $p(x) = 0$ if and only if x is in C and then, for each positive integer k, we minimize

$$P_k(x) = f(x) + kp(x), \tag{2.3}$$

to get x^k. We then want the sequence $\{x^k\}$ to converge to some $x^* \in C$ that solves the original problem. In order for this iterative algorithm to be useful, each x^k should be relatively easy to calculate.

If, for example, we should select $p(x) = +\infty$ for x not in C and $p(x) = 0$ for x in C, then minimizing $P_k(x)$ is equivalent to the original problem and we have achieved nothing.

As an example, suppose that we want to minimize the function $f(x) = (x+1)^2$, subject to $x \geq 0$. Let us select $p(x) = x^2$, for $x \leq 0$, and $p(x) = 0$ otherwise. Then $x^k = \frac{-1}{k+1}$, which converges to the right answer, $x^* = 0$, as $k \to \infty$.

2.3 Auxiliary-Function Methods

In this section we define auxiliary-function methods, establish their basic properties, and give several examples to be considered in more detail later.

2.3.1 General AF Methods

Let C be a nonempty subset of an arbitrary set X, and $f : X \to \mathbb{R}$. We want to minimize $f(x)$ over x in C. At the kth step of an auxiliary-function (AF) algorithm we minimize

$$G_k(x) = f(x) + g_k(x) \tag{2.4}$$

over $x \in C$ to obtain x^k. Our main objective is to select the $g_k(x)$ so that the infinite sequence $\{x^k\}$ generated by our algorithm converges to a solution of the problem; this, of course, requires some topology on the set X. Failing that, we want the sequence $\{f(x^k)\}$ to converge to $d = \inf\{f(x)|x \in C\}$ or, at the very least, for the sequence $\{f(x^k)\}$ to be nonincreasing.

2.3.2 AF Requirements

For AF methods we require that the auxiliary functions $g_k(x)$ be chosen so that $g_k(x) \geq 0$ for all $x \in C$ and $g_k(x^{k-1}) = 0$. We then have the following proposition.

Proposition 2.1 *Let the sequence $\{x^k\}$ be generated by an AF algorithm. Then the sequence $\{f(x^k)\}$ is nonincreasing, and, if d is finite, the sequence $\{g_k(x^k)\}$ converges to zero.*

Proof: We have

$$f(x^k) + g_k(x^k) = G_k(x^k) \leq G_k(x^{k-1}) = f(x^{k-1}) + g_k(x^{k-1}) = f(x^{k-1}).$$

Therefore,

$$f(x^{k-1}) - f(x^k) \geq g_k(x^k) \geq 0.$$

Since the sequence $\{f(x^k)\}$ is decreasing and bounded below by d, the difference sequence must converge to zero, if d is finite; therefore, the sequence $\{g_k(x^k)\}$ converges to zero in this case. ∎

The auxiliary functions used in Equation (2.2) do not have these properties but the barrier-function algorithm can be reformulated as an AF method. The iterate x^k obtained by minimizing $B_k(x)$ in Equation (2.2) also minimizes the function

$$G_k(x) = f(x) + [(k-1)f(x) + b(x)] - [(k-1)f(x^{k-1}) + b(x^{k-1})]. \quad (2.5)$$

The auxiliary functions

$$g_k(x) = [(k-1)f(x) + b(x)] - [(k-1)f(x^{k-1}) + b(x^{k-1})] \quad (2.6)$$

now have the desired properties. In addition, we have $G_k(x) - G_k(x^k) = g_{k+1}(x)$ for all $x \in C$, which will become significant shortly.

As originally formulated, the penalty-function methods do not fit into the class of AF methods we consider here. However, a reformulation of the penalty-function approach, with $p(x)$ and $f(x)$ switching roles, permits the penalty-function methods to be studied as barrier-function methods, and therefore as acceptable AF methods.

2.3.3 Majorization Minimization

Majorization minimization (MM), also called optimization transfer, is a technique used in statistics to convert a difficult optimization problem into a sequence of simpler ones [183, 21, 157]. The MM method requires that we majorize the objective function $f(x)$ with $g(x|y)$, such that $g(x|y) \geq f(x)$, for all x, and $g(y|y) = f(y)$. At the kth step of the iterative algorithm we minimize the function $g(x|x^{k-1})$ to get x^k.

The MM methods are members of the AF class. At the kth step of an MM iteration we minimize

$$G_k(x) = f(x) + [g(x|x^{k-1}) - f(x)] = f(x) + d(x, x^{k-1}), \quad (2.7)$$

where $d(x,z)$ is some distance function satisfying $d(x,z) \geq 0$ and $d(z,z) = 0$. Since $g_k(x) = d(x, x^{k-1}) \geq 0$ and $g_k(x^{k-1}) = 0$, MM methods are also AF methods; it then follows that the sequence $\{f(x^k)\}$ is nonincreasing.

All MM algorithms have the form $x^k = Tx^{k-1}$, where T is the operator defined by

$$Tz = \text{argmin}_x \{f(x) + d(x,z)\}. \quad (2.8)$$

If $d(x,z) = \frac{1}{2}\|x-z\|_2^2$, then T is Moreau's proximity operator $Tz = \text{prox}_f(z)$ [169, 170, 171].

2.3.4 The Method of Auslander and Teboulle

The method of Auslander and Teboulle [7] is a particular example of an MM algorithm. We take C to be a closed, nonempty, convex subset of $\mathbb{R}^J$, with interior U. At the kth step of their method one minimizes a function

$$G_k(x) = f(x) + d(x, x^{k-1}) \tag{2.9}$$

to get x^k. Their distance $d(x, y)$ is defined for x and y in U, and the gradient with respect to the first variable, denoted $\nabla_1 d(x, y)$, is assumed to exist. The distance $d(x, y)$ is not assumed to be a Bregman distance. Instead, they assume that the distance d has an associated *induced proximal distance* $H(a, b) \geq 0$, finite for a and b in U, with $H(a, a) = 0$ and

$$\langle \nabla_1 d(b, a), c - b \rangle \leq H(c, a) - H(c, b), \tag{2.10}$$

for all c in U.

If $d = D_h$, that is, if d is a Bregman distance, then from the equation

$$\langle \nabla_1 d(b, a), c - b \rangle = D_h(c, a) - D_h(c, b) - D_h(b, a) \tag{2.11}$$

we see that D_h has $H = D_h$ for its associated induced proximal distance, so D_h is *self-proximal*, in the terminology of [7].

2.3.5 The EM Algorithm

The *expectation maximization maximum likelihood* (EM) "algorithm" is not a single algorithm, but a framework, or, as the authors of [21] put it, a "prescription", for constructing algorithms. Nevertheless, we shall refer to it as the EM algorithm.

The EM algorithm is always presented within the context of statistical likelihood maximization, but the essence of this method is not stochastic; the EM algorithms can be shown to form a subclass of AF methods. We present now the essential aspects of the EM algorithm without relying on statistical concepts.

The problem is to maximize a nonnegative function $f : Z \to \mathbb{R}$, where Z is an arbitrary set; in the stochastic context $f(z)$ is a likelihood function of the parameter vector z. We assume that there is $z^* \in Z$ with $f(z^*) \geq f(z)$, for all $z \in Z$.

We also assume that there is a nonnegative function $b : \mathbb{R}^J \times Z \to \mathbb{R}$ such that

$$f(z) = \int b(x, z) dx.$$

Having found z^{k-1}, we maximize the function

$$H(z^{k-1}, z) = \int b(x, z^{k-1}) \log b(x, z) dx \tag{2.12}$$

to get z^k. Adopting such an iterative approach presupposes that maximizing $H(z^{k-1}, z)$ is simpler than maximizing $f(z)$ itself. This is the case with the EM algorithms.

One of the most useful and easily proved facts about the Kullback-Leibler distance is contained in the following lemma.

Lemma 2.1 *For nonnegative vectors x and z, with $z_+ = \sum_{j=1}^{J} z_j > 0$, we have*

$$KL(x, z) = KL(x_+, z_+) + KL(x, \frac{x_+}{z_+} z). \tag{2.13}$$

This lemma can be extended to obtain the following useful identity; we simplify the notation by setting $b(z) = b(x, z)$.

Lemma 2.2 *For $f(z)$ and $b(x, z)$ as above, and z and w in Z, with $f(w) > 0$, we have*

$$KL(b(z), b(w)) = KL(f(z), f(w)) + KL(b(z), (f(z)/f(w))b(w)). \tag{2.14}$$

Maximizing $H(z^{k-1}, z)$ is equivalent to minimizing

$$G_k(z) = G(z^{k-1}, z) = -f(z) + KL(b(z^{k-1}), b(z)), \tag{2.15}$$

where

$$g_k(z) = KL(b(z^{k-1}), b(z)) = \int KL(b(x, z^{k-1}), b(x, z))dx. \tag{2.16}$$

Since $g_k(z) \geq 0$ for all z and $g_k(z^{k-1}) = 0$, we have an AF method. Without additional restrictions, we cannot conclude that $\{f(z^k)\}$ converges to $f(z^*)$.

We get z^k by minimizing $G_k(z) = G(z^{k-1}, z)$. When we minimize $G(z, z^k)$, we get z^k again. Therefore, we can put the EM algorithm into the alternating minimization (AM) framework of Csiszár and Tusnády [105], to be discussed later.

2.4 The SUMMA Class of AF Methods

As we have seen, whenever the sequence $\{x^k\}$ is generated by an AF algorithm, the sequence $\{f(x^k)\}$ is nonincreasing. We want more, however; we want the sequence $\{f(x^k)\}$ to converge to $d = \inf_{x \in C} f(x)$. This happens for those AF algorithms in the SUMMA class.

2.4.1 The SUMMA Property

An AF algorithm is said to be in the SUMMA class if the auxiliary functions $g_k(x)$ are chosen so that the SUMMA property holds; that is,

$$G_k(x) - G_k(x^k) \geq g_{k+1}(x) \geq 0, \tag{2.17}$$

for all $x \in C$. As we saw previously, the reformulated barrier-function method is in the SUMMA class. We have the following theorem.

Theorem 2.1 *If the sequence $\{x^k\}$ is generated by an algorithm in the SUMMA class, then the sequence $\{f(x^k)\}$ converges to $d = \inf_{x \in C} f(x)$.*

Proof: Suppose that there is $d^* > d$ with $f(x^k) \geq d^*$, for all k. Then there is z in C with

$$f(x^k) \geq d^* > f(z) \geq d,$$

for all k. Using the inequality (2.17) we have

$$g_k(z) - g_{k+1}(z) \geq f(x^k) + g_k(x^k) - f(z) \geq f(x^k) - f(z) \geq d^* - f(z) > 0.$$

This tells us that the nonnegative sequence $\{g_k(z)\}$ is decreasing, but that successive differences remain bounded away from zero, which cannot happen. ∎

2.4.2 Auslander and Teboulle Revisited

The method of Auslander and Teboulle described previously seems not to be a particular case of SUMMA. However, we can adapt the proof of Theorem 2.1 to prove the analogous result for their method. We assume that $f(\hat{x}) \leq f(x)$, for all x in C.

Theorem 2.2 *For $k = 2, 3, ...$, let x^k minimize the function*

$$G_k(x) = f(x) + d(x, x^{k-1}).$$

If the distance d has an induced proximal distance H, then $\{f(x^k)\} \to f(\hat{x})$.

Proof: We know that the sequence $\{f(x^k)\}$ is decreasing and the sequence $\{d(x^k, x^{k-1})\}$ converges to zero. Now suppose that

$$f(x^k) \geq f(\hat{x}) + \delta,$$

for some $\delta > 0$ and all k. Since $\hat{x}$ is in C, there is z in U with

$$f(x^k) \geq f(z) + \frac{\delta}{2},$$

for all k. Since x^k minimizes $F_k(x)$, it follows that

$$0 = \nabla f(x^k) + \nabla_1 d(x^k, x^{k-1}).$$

Using the convexity of the function $f(x)$ and the fact that H is an induced proximal distance, we have

$$0 < \frac{\delta}{2} \leq f(x^k) - f(z) \leq \langle -\nabla f(x^k), z - x^k \rangle =$$

$$\langle \nabla_1 d(x^k, x^{k-1}), z - x^k \rangle \le H(z, x^{k-1}) - H(z, x^k).$$

Therefore, the nonnegative sequence $\{H(z, x^k)\}$ is decreasing, but its successive differences remain bounded below by $\frac{\delta}{2}$, which is a contradiction. ∎

It is interesting to note that the Auslander-Teboulle approach places a restriction on the function $d(x, y)$, the existence of the induced proximal distance H, that is unrelated to the objective function $f(x)$, but this condition is helpful only for convex $f(x)$. In contrast, the SUMMA approach requires that

$$0 \le g_{k+1}(x) \le G_k(x) - G_k(x^k),$$

which involves the $f(x)$ being minimized, but does not require that this $f(x)$ be convex.

2.4.3 Proximal Minimization

Let $f : \mathbb{R}^J \to (-\infty, +\infty]$ be a convex function. Let h be another convex function, with effective domain D, that is differentiable on the nonempty open convex set int D. Assume that $f(x)$ is finite on $C = \overline{D}$ and attains its minimum value on C at $\hat{x}$. The corresponding *Bregman distance* [31] $D_h(x, z)$ is defined for x in D and z in int D by

$$D_h(x, z) = h(x) - h(z) - \langle \nabla h(z), x - z \rangle. \tag{2.18}$$

Note that $D_h(x, z) \ge 0$ always. If h is essentially strictly convex, then $D_h(x, z) = 0$ implies that $x = z$. Our objective is to minimize $f(x)$ over x in $C = \overline{D}$.

At the kth step of a *proximal minimization algorithm* (PMA) [95, 55], we minimize the function

$$G_k(x) = f(x) + D_h(x, x^{k-1}), \tag{2.19}$$

to get x^k. The function

$$g_k(x) = D_h(x, x^{k-1}) \tag{2.20}$$

is nonnegative and $g_k(x^{k-1}) = 0$. We assume that each x^k lies in int D. As we shall see,

$$G_k(x) - G_k(x^k) \ge D_h(x, x^k) = g_{k+1}(x) \ge 0, \tag{2.21}$$

so the PMA is in the SUMMA class.

The PMA can present some computational obstacles. When we minimize $G_k(x)$ to get x^k we find that we must solve the equation

$$\nabla h(x^{k-1}) - \nabla h(x^k) \in \partial f(x^k), \tag{2.22}$$

where the set $\partial f(x)$ is the sub-differential of f at x, given by

$$\partial f(x) := \{u | \langle u, y - x \rangle \leq f(y) - f(x), \text{for all}\, y\}. \tag{2.23}$$

When $f(x)$ is differentiable, we must solve

$$\nabla f(x^k) + \nabla h(x^k) = \nabla h(x^{k-1}). \tag{2.24}$$

A modification of the PMA, called the IPA for *interior-point algorithm* [55, 63], is designed to overcome these computational obstacles. We describe the IPA in the next subsection. Another modification of the PMA that is similar to the IPA is the *forward-backward splitting* (FBS) method to be discussed in Chapter 5.

2.4.4 The IPA

In this subsection we describe a modification of the PMA, an interior-point algorithm called the IPA, that helps us overcome the computational obstacles encountered in the PMA. To simplify the discussion, we assume in this subsection that $f(x)$ is differentiable.

At the kth step of the PMA we minimize

$$G_k(x) = f(x) + D_h(x, x^{k-1}), \tag{2.25}$$

where $h(x)$ is as in the previous subsection. Writing

$$a(x) = h(x) + f(x), \tag{2.26}$$

we must solve the equation

$$\nabla a(x^k) = \nabla a(x^{k-1}) - \nabla f(x^{k-1}). \tag{2.27}$$

In the IPA we select $a(x)$ so that Equation (2.27) is easily solved and so that $h(x) = a(x) - f(x)$ is convex and differentiable. We shall present several examples of the IPA in Chapter 4.

Chapter 3

Barrier-Function and Penalty-Function Methods

Barrier-function and penalty-function methods are the best known examples of sequential optimization. In their usual formulations neither fits into the AF class of algorithms. However, barrier-function algorithms can be reformulated to fit into the SUMMA class, while penalty-function methods can be converted to barrier-function methods by switching the roles of the objective and penalty functions.

3.1 Barrier Functions

Let $b(x) : \mathbb{R}^J \to (-\infty, +\infty]$ be continuous, with effective domain the set

$$D = \{x|\, b(x) < +\infty\}.$$

The goal is to minimize the objective function $f(x)$, over x in C, the closure of D. We assume that there is $\hat{x} \in C$ with $f(\hat{x}) \leq f(x)$, for all x in C.

In the barrier-function method, we minimize

$$B_k(x) = f(x) + \frac{1}{k} b(x) \tag{3.1}$$

over x in D to get x^k. Each x^k lies within D, so the method is an interior-point algorithm. If the sequence $\{x^k\}$ converges, the limit vector x^* will be in C and $f(x^*) = f(\hat{x})$.

Barrier functions typically have the property that $b(x) \to +\infty$ as x approaches the boundary of D, so not only is x^k prevented from leaving D, it is discouraged from approaching the boundary.

3.2 Examples of Barrier Functions

Consider the convex programming (CP) problem of minimizing the convex function $f : \mathbb{R}^J \to \mathbb{R}$, subject to $g_i(x) \leq 0$, where each $g_i : \mathbb{R}^J \to \mathbb{R}$ is convex, for $i = 1, ..., I$. Let $D = \{x | g_i(x) < 0, i = 1, ..., I\}$; then D is open. We consider two barrier functions appropriate for this problem.

3.2.1 The Logarithmic Barrier Function

A suitable barrier function is the *logarithmic barrier function*

$$b(x) = \Big(-\sum_{i=1}^{I} \log(-g_i(x)) \Big). \tag{3.2}$$

The function $-\log(-g_i(x))$ is defined only for those x in D, and is positive for $g_i(x) > -1$. If $g_i(x)$ is near zero, then so is $-g_i(x)$ and $b(x)$ will be large.

3.2.2 The Inverse Barrier Function

Another suitable barrier function is the *inverse barrier function*

$$b(x) = \sum_{i=1}^{I} \frac{-1}{g_i(x)}, \tag{3.3}$$

defined for those x in D.

In both examples, when k is small, the minimization pays more attention to $b(x)$, and less to $f(x)$, forcing the $g_i(x)$ to be large negative numbers.

But, as k grows larger, more attention is paid to minimizing $f(x)$ and the $g_i(x)$ are allowed to be smaller negative numbers. By letting $k \to \infty$, we obtain an iterative method for solving the constrained minimization problem.

Barrier-function methods are particular cases of the SUMMA. The iterative step of the barrier-function method can be formulated as follows: minimize

$$f(x) + [(k-1)f(x) + b(x)] \tag{3.4}$$

to get x^k. Since, for $k = 2, 3, \ldots$, the function

$$(k-1)f(x) + b(x) \tag{3.5}$$

is minimized by x^{k-1}, the function

$$g_k(x) = (k-1)f(x) + b(x) - (k-1)f(x^{k-1}) - b(x^{k-1}) \tag{3.6}$$

is nonnegative, and x^k minimizes the function

$$G_k(x) = f(x) + g_k(x). \tag{3.7}$$

From

$$G_k(x) = f(x) + (k-1)f(x) + b(x) - (k-1)f(x^{k-1}) - b(x^{k-1}),$$

it follows that

$$G_k(x) - G_k(x^k) = kf(x) + b(x) - kf(x^k) - b(x^k) = g_{k+1}(x),$$

so that $g_{k+1}(x)$ satisfies the condition in (2.17). This shows that the barrier-function method is a particular case of SUMMA.

From the properties of SUMMA algorithms, we conclude that $\{f(x^k)\}$ is decreasing to $f(\hat{x})$, and that $\{g_k(x^k)\}$ converges to zero. From the nonnegativity of $g_k(x^k)$ we have that

$$(k-1)(f(x^k) - f(x^{k-1})) \geq b(x^{k-1}) - b(x^k).$$

Since the sequence $\{f(x^k)\}$ is decreasing, the sequence $\{b(x^k)\}$ must be increasing, but might not be bounded above.

If $\hat{x}$ is unique, and $f(x)$ has bounded level sets, then it follows, from our discussion of SUMMA, that $\{x^k\} \to \hat{x}$. Suppose now that $\hat{x}$ is not known to be unique, but can be chosen in D, so that $G_k(\hat{x})$ is finite for each k. From

$$f(\hat{x}) + \frac{1}{k}b(\hat{x}) \geq f(x^k) + \frac{1}{k}b(x^k)$$

we have

$$\frac{1}{k}\Big(b(\hat{x}) - b(x^k)\Big) \geq f(x^k) - f(\hat{x}) \geq 0,$$

so that

$$b(\hat{x}) - b(x^k) \geq 0,$$

for all k. If either f or b has bounded level sets, then the sequence $\{x^k\}$ is bounded and has a cluster point, x^* in C. It follows that $b(x^*) \leq b(\hat{x}) < +\infty$, so that x^* is in D. If we assume that $f(x)$ is convex and $b(x)$ is strictly convex on D, then we can show that x^* is unique in D, so that $x^* = \hat{x}$ and $\{x^k\} \to \hat{x}$.

To see this, assume, to the contrary, that there are two distinct cluster points x^* and x^{**} in D, with

$$\{x^{k_n}\} \to x^*,$$

and

$$\{x^{j_n}\} \to x^{**}.$$

Without loss of generality, we assume that

$$0 < k_n < j_n < k_{n+1},$$

for all n, so that

$$b(x^{k_n}) \leq b(x^{j_n}) \leq b(x^{k_{n+1}}).$$

Therefore,

$$b(x^*) = b(x^{**}) \leq b(\hat{x}).$$

From the strict convexity of $b(x)$ on the set D, and the convexity of $f(x)$, we conclude that, for $0 < \lambda < 1$ and $y = (1-\lambda)x^* + \lambda x^{**}$, we have $b(y) < b(x^*)$ and $f(y) \leq f(x^*)$. But, we must then have $f(y) = f(x^*)$. There must then be some k_n such that

$$G_{k_n}(y) = f(y) + \frac{1}{k_n} b(y) < f(x_{k_n}) + \frac{1}{k_n} b(x_{k_n}) = G_{k_n}(x^{k_n}).$$

But, this is a contradiction. ∎

The following theorem summarizes what we have shown with regard to the barrier-function method.

Theorem 3.1 *Let $f : \mathbb{R}^J \to (-\infty, +\infty]$ be a continuous function. Let $b(x) : \mathbb{R}^J \to (0, +\infty]$ be a continuous function, with effective domain the nonempty set D. Let $\hat{x}$ minimize $f(x)$ over all x in $C = \overline{D}$. For each positive integer k, let x^k minimize the function $f(x) + \frac{1}{k} b(x)$. Then the sequence $\{f(x^k)\}$ is monotonically decreasing to the limit $f(\hat{x})$, and the sequence $\{b(x^k)\}$ is increasing. If $\hat{x}$ is unique, and $f(x)$ has bounded level sets, then the sequence $\{x^k\}$ converges to $\hat{x}$. In particular, if $\hat{x}$ can be chosen in D, if either $f(x)$ or $b(x)$ has bounded level sets, if $f(x)$ is convex and if $b(x)$ is strictly convex on D, then $\hat{x}$ is unique in D and $\{x^k\}$ converges to $\hat{x}$.*

At the kth step of the barrier method we must minimize the function $f(x) + \frac{1}{k}b(x)$. In practice, this must also be performed iteratively, with, say, the Newton-Raphson algorithm. It is important, therefore, that barrier functions be selected so that relatively few Newton-Raphson steps are needed to produce acceptable solutions to the main problem. For more on these issues see Renegar [191] and Nesterov and Nemirovski [176].

3.3 Penalty Functions

When we add a barrier function to $f(x)$ we restrict the domain. When the barrier function is used in a sequential unconstrained minimization algorithm, the vector x^k that minimizes the function $B_k(x) = f(x) + \frac{1}{k}b(x)$ lies in the effective domain D of $b(x)$, and we proved that, under certain conditions, the sequence $\{x^k\}$ converges to a minimizer of the function $f(x)$ over the closure of D. The constraint of lying within the set $\overline{D}$ is satisfied at every step of the algorithm; for that reason such algorithms are called interior-point methods. Constraints may also be imposed using a penalty function. In this case, violations of the constraints are discouraged, but not forbidden. When a penalty function is used in a sequential unconstrained minimization algorithm, the x^k need not satisfy the constraints; only the limit vector need be feasible.

3.4 Examples of Penalty Functions

Consider the convex programming problem. We wish to minimize the convex function $f(x)$ over all x for which the convex functions $g_i(x) \leq 0$, for $i = 1, ..., I$.

3.4.1 The Absolute-Value Penalty Function

We let $g_i^+(x) = \max\{g_i(x), 0\}$, and

$$p(x) = \sum_{i=1}^{I} g_i^+(x). \tag{3.8}$$

This is the *Absolute-Value* penalty function; it penalizes violations of the constraints $g_i(x) \leq 0$, but does not forbid such violations. Then, for $k =$

$1, 2, ...,$ we minimize

$$P_k(x) = f(x) + kp(x), \tag{3.9}$$

to get x^k. As $k \to +\infty$, the penalty function becomes more heavily weighted, so that, in the limit, the constraints $g_i(x) \le 0$ should hold. Because only the limit vector satisfies the constraints, and the x^k are allowed to violate them, such a method is called an *exterior-point* method.

3.4.2 The Courant-Beltrami Penalty Function

The *Courant-Beltrami* penalty-function method is similar, but uses

$$p(x) = \sum_{i=1}^{I} [g_i^+(x)]^2. \tag{3.10}$$

3.4.3 The Quadratic-Loss Penalty Function

Penalty methods can also be used with equality constraints. Consider the problem of minimizing the convex function $f(x)$, subject to the constraints $g_i(x) = 0$, $i = 1, ..., I$. The *quadratic-loss* penalty function is

$$p(x) = \frac{1}{2}\sum_{i=1}^{I} (g_i(x))^2. \tag{3.11}$$

The inclusion of a penalty term can serve purposes other than to impose constraints on the location of the limit vector. In image processing, it is often desirable to obtain a reconstructed image that is locally smooth, but with well defined edges. Penalty functions that favor such images can then be used in the iterative reconstruction [128]. We survey several instances in which we would want to use a penalized objective function.

3.4.4 Regularized Least-Squares

Suppose we want to solve the system of equations $Ax = b$. The problem may have no exact solution, precisely one solution, or there may be infinitely many solutions. If we minimize the function

$$f(x) = \frac{1}{2}\|Ax - b\|_2^2,$$

we get a *least-squares* solution, generally, and an exact solution, whenever exact solutions exist. When the matrix A is ill-conditioned, small changes in the vector b can lead to large changes in the solution. When the vector

b comes from measured data, the entries of b may include measurement errors, so that an exact solution of $Ax = b$ may be undesirable, even when such exact solutions exist; exact solutions may correspond to x with unacceptably large norm, for example. In such cases, we may, instead, wish to minimize a function such as

$$\frac{1}{2}\|Ax - b\|_2^2 + \frac{\epsilon}{2}\|x - z\|_2^2, \tag{3.12}$$

for some vector z. If $z = 0$, the minimizing vector x_ϵ is then a *norm-constrained* least-squares solution. We then say that the least-squares problem has been *regularized*. In the limit, as $\epsilon \to 0$, these regularized solutions x_ϵ converge to the least-squares solution closest to z.

Suppose the system $Ax = b$ has infinitely many exact solutions. Our problem is to select one. Let us select z that incorporates features of the desired solution, to the extent that we know them *a priori*. Then, as $\epsilon \to 0$, the vectors x_ϵ converge to the exact solution closest to z. For example, taking $z = 0$ leads to the *minimum-norm solution*.

3.4.5 Minimizing Cross-Entropy

In image processing, it is common to encounter systems $Px = y$ in which all the terms are nonnegative. In such cases, it may be desirable to solve the system $Px = y$, approximately, perhaps, by minimizing the *cross-entropy* or *Kullback-Leibler distance*

$$KL(y, Px) = \sum_{i=1}^{I} \Big(y_i \log \frac{y_i}{(Px)_i} + (Px)_i - y_i \Big), \tag{3.13}$$

over vectors $x \geq 0$. When the vector y is noisy, the resulting solution, viewed as an image, can be unacceptable. It is wise, therefore, to add a penalty term, such as $p(x) = \epsilon KL(z, x)$, where $z > 0$ is a prior estimate of the desired x [155, 206, 156, 47].

A similar problem involves minimizing the function $KL(Px, y)$. Once again, noisy results can be avoided by including a penalty term, such as $p(x) = \epsilon KL(x, z)$ [47].

3.4.6 The Lagrangian in Convex Programming

When there is a sensitivity vector λ for the CP problem, minimizing $f(x)$ is equivalent to minimizing the Lagrangian,

$$f(x) + \sum_{i=1}^{I} \lambda_i g_i(x) = f(x) + p(x); \tag{3.14}$$

in this case, the addition of the second term, $p(x)$, serves to incorporate the constraints $g_i(x) \leq 0$ in the function to be minimized, turning a constrained minimization problem into an unconstrained one. The problem of minimizing the Lagrangian still remains, though. We may have to solve that problem using an iterative algorithm.

3.4.7 Infimal Convolution

The *infimal convolution* of the functions f and g is defined as

$$(f \oplus g)(z) = \inf_x \Big\{ f(x) + g(z - x) \Big\}.$$

The *infimal deconvolution* of f and g is defined as

$$(f \ominus g)(z) = \sup_x \Big\{ f(z - x) - g(x) \Big\}.$$

3.4.8 Moreau's Proximity-Function Method

The Moreau envelope of the function f is the function

$$m_f(z) = \inf_x \Big\{ f(x) + \frac{1}{2}\|x - z\|_2^2 \Big\}, \tag{3.15}$$

which is also the *infimal convolution* of the functions $f(x)$ and $\frac{1}{2}\|x\|_2^2$. It can be shown that the infimum is uniquely attained at the point denoted $x = \text{prox}_f z$ (see [192]). In similar fashion, we can define $m_{f^*} z$ and $\text{prox}_{f^*} z$, where $f^*(z)$ denotes the function conjugate to f.

Let z be fixed and $\hat{x}$ minimize the function

$$f(x) + \frac{1}{2\gamma}\|x - z\|_2^2. \tag{3.16}$$

Then we have

$$0 \in \partial f(\hat{x}) + \frac{1}{\gamma}(\hat{x} - z),$$

or

$$z - \hat{x} \in \partial(\gamma f)(\hat{x}).$$

If $z - x \in \partial f(x)$ and $z - y \in \partial f(y)$, then $x = y$: we have

$$f(y) - f(x) \geq \langle z - x, y - x \rangle,$$

and

$$f(x) - f(y) \geq \langle z - y, x - y \rangle = -\langle z - y, y - x \rangle.$$

Adding, we get

$$0 \geq \langle y - x, y - x \rangle = \|x - y\|_2^2.$$

We can then say that $x = \text{prox}_f(z)$ is characterized by the inequality

$$z - x \in \partial f(x). \tag{3.17}$$

Consequently, we can write

$$\hat{x} = \text{prox}_{\gamma f}(z).$$

Proposition 3.1 *The infimum of $m_f(z)$, over all z, is the same as the infimum of $f(x)$, over all x.*

Proof: We have

$$\inf_z m_f(z) = \inf_z \inf_x \{f(x) + \frac{1}{2}\|x - z\|_2^2\}$$

$$= \inf_x \inf_z \{f(x) + \frac{1}{2}\|x - z\|_2^2\} = \inf_x \{f(x) + \frac{1}{2} \inf_z \|x - z\|_2^2\} = \inf_x f(x).$$

∎

The minimizers of $m_f(z)$ and $f(x)$ are the same, as well. Therefore, one way to use Moreau's method is to replace the original problem of minimizing the possibly non-smooth function $f(x)$ with the problem of minimizing the smooth function $m_f(z)$. Another way is to convert Moreau's method into a sequential minimization algorithm, replacing z with x^{k-1} and minimizing with respect to x to get x^k. This leads to the proximal minimization algorithm.

3.5 Basic Facts

Once again, our objective is to find a sequence $\{x^k\}$ such that $\{f(x^k)\} \to d$. We select a penalty function $p(x)$ with $p(x) \geq 0$ and $p(x) = 0$ if and only if x is in C. For $k = 1, 2, ...$, let x^k be a minimizer of the function $f(x) + kp(x)$. As we shall see, we can formulate this penalty-function algorithm as a barrier-function iteration.

In order to relate penalty-function methods to barrier-function methods, we note that minimizing $P_k(x) = f(x) + kp(x)$ is equivalent to minimizing $p(x) + \frac{1}{k} f(x)$. This is the form of the barrier-function iteration, with $p(x)$ now in the role previously played by $f(x)$, and $f(x)$ now in the role previously played by $b(x)$. We are not concerned here with the effective domain of $f(x)$. Therefore, we can now mimic most, but not all, of what we did for barrier-function methods.

We assume that there is $\alpha \in \mathbb{R}$ such that $\alpha \leq f(x)$, for all $x \in \mathbb{R}^J$.

Lemma 3.1 *The sequence $\{P_k(x^k)\}$ is increasing, bounded above by d and converges to some $\gamma \leq d$.*

Proof: We have

$$P_k(x^k) \leq P_k(x^{k+1}) \leq P_k(x^{k+1}) + p(x^{k+1}) = T_{k+1}(x^{k+1}).$$

Also, for any $z \in C$, and for each k, we have

$$f(z) = f(z) + kp(z) = P_k(z) \geq P_k(x^k);$$

therefore $d \geq \gamma$. ∎

Lemma 3.2 *The sequence $\{p(x^k)\}$ is decreasing to zero, the sequence $\{f(x^k)\}$ is increasing and converging to some $\beta \leq d$.*

Proof: Since x^k minimizes $P_k(x)$ and x^{k+1} minimizes $T_{k+1}(x)$, we have

$$f(x^k) + kp(x^k) \leq f(x^{k+1}) + kp(x^{k+1}),$$

and

$$f(x^{k+1}) + (k+1)p(x^{k+1}) \leq f(x^k) + (k+1)p(x^k).$$

Consequently, we have

$$(k+1)[p(x^k) - p(x^{k+1})] \geq f(x^{k+1}) - f(x^k) \geq k[p(x^k) - p(x^{k+1})].$$

Therefore,

$$p(x^k) - p(x^{k+1}) \geq 0,$$

and

$$f(x^{k+1}) - f(x^k) \geq 0.$$

From

$$f(x^k) \leq f(x^k) + kp(x^k) = P_k(x^k) \leq \gamma \leq d,$$

it follows that the sequence $\{f(x^k)\}$ is increasing and converges to some $\beta \leq \gamma$. Since

$$\alpha + kp(x^k) \leq f(x^k) + kp(x^k) = P_k(x^k) \leq \gamma$$

for all k, we have $0 \leq kp(x^k) \leq \gamma - \alpha$. Therefore, the sequence $\{p(x^k)\}$ converges to zero. ∎

We want $\beta = d$. To obtain this result, it appears that we need to make more assumptions: we assume, therefore, that X is a complete metric space, C is closed in X, the functions f and p are continuous and f has compact level sets. From these assumptions, we are able to assert that the sequence

$\{x^k\}$ is bounded, so that there is a convergent subsequence; let $\{x^{k_n}\} \to x^*$. It follows that $p(x^*) = 0$, so that x^* is in C. Then

$$f(x^*) = f(x^*) + p(x^*) = \lim_{n\to+\infty} (f(x^{k_n}) + p(x^{k_n})) \leq \lim_{n\to+\infty} T_{k_n}(x^{k_n}) = \gamma \leq d.$$

But $x^* \in C$, so $f(x^*) \geq d$. Therefore, $f(x^*) = d$.

It may seem odd that we are trying to minimize $f(x)$ over the set C using a sequence $\{x^k\}$ with $\{f(x^k)\}$ increasing, but remember that these x^k are not in C.

Definition 3.1 *Let X be a complete metric space. A real-valued function $p(x)$ on X has* compact level sets *if, for all real γ, the level set $\{x|p(x) \leq \gamma\}$ is compact.*

Theorem 3.2 *Let X be a complete metric space, $f(x)$ be a continuous function, and the restriction of $f(x)$ to x in C have compact level sets. Then the sequence $\{x^k\}$ is bounded and has convergent subsequences. Furthermore, $f(x^*) = d$, for any subsequential limit point $x^* \in X$. If $\hat{x}$ is the unique minimizer of $f(x)$ for $x \in C$, then $x^* = \hat{x}$ and $\{x^k\} \to \hat{x}$.*

Proof: From the previous theorem we have $f(x^*) = d$, for all subsequential limit points x^*. But, by uniqueness, $x^* = \hat{x}$, and so $\{x^k\} \to \hat{x}$. ∎

Corollary 3.1 *Let $C \subseteq \mathbb{R}^J$ be closed and convex. Let $f(x) : \mathbb{R}^J \to \mathbb{R}$ be closed, proper and convex. If $\hat{x}$ is the unique minimizer of $f(x)$ over $x \in C$, the sequence $\{x^k\}$ converges to $\hat{x}$.*

Proof: Let $\iota_C(x)$ be the indicator function of the set C, that is, $\iota_C(x) = 0$, for all x in C, and $\iota_C(x) = +\infty$, otherwise. Then the function $g(x) = f(x) + \iota_C(x)$ is closed, proper and convex. If $\hat{x}$ is unique, then we have

$$\{x|f(x) + \iota_C(x) \leq f(\hat{x})\} = \{\hat{x}\}.$$

Therefore, one of the level sets of $g(x)$ is bounded and nonempty. It follows from Corollary 8.7.1 of [192] that every level set of $g(x)$ is bounded, so that the sequence $\{x^k\}$ is bounded. ∎

If $\hat{x}$ is not unique, we can still prove convergence of the sequence $\{x^k\}$ for particular cases of SUMMA.

Chapter 4

Proximal Minimization

In this chapter we consider the use of Bregman distances in constrained optimization through the *proximal minimization* method. The *proximal minimization algorithm* (PMA) is in the SUMMA class and this fact is used to establish important properties of the PMA. A detailed discussion of the PMA and its history is found in the book by Censor and Zenios [95].

4.1 The Basic Problem

We want to minimize a convex function $f : \mathbb{R}^J \to \mathbb{R}$ over a closed, nonempty convex subset $C \subseteq \mathbb{R}^J$. If the problem is ill-conditioned in some way, perhaps because the function $f(x)$ is not strictly convex, then regularization is needed.

For example, the least-squares approximate solution of $Ax = b$ is obtained by minimizing the function $f(x) = \frac{1}{2}\|Ax - b\|_2^2$ over all x. When the matrix A is ill-conditioned the least-squares solution may have a large two-norm. To regularize the least-squares problem we can impose a norm constraint and minimize

$$\frac{1}{2}\|Ax-b\|_2^2+\frac{\epsilon}{2}\|x\|_2^2, \tag{4.1}$$

where $\epsilon > 0$ is small.

Returning to our original problem, we can impose strict convexity and regularize by minimizing the function

$$f(x)+\frac{1}{2k}\|x-a\|_2^2 \tag{4.2}$$

to get x^k, for some selected vector a and $k = 1, 2, ...$ One difficulty with this approach is that, for small k, there may be too much emphasis on the second term in Equation (4.2), while the problem becomes increasingly ill-conditioned as k increases. As pointed out in [95], one way out of this difficulty is to obtain x^k by minimizing

$$f(x)+\frac{\gamma}{2}\|x-x^{k-1}\|_2^2. \tag{4.3}$$

This suggests a more general technique for constrained optimization, called *proximal minimization* with D-functions in [95].

4.2 Proximal Minimization Algorithms

Let $f : \mathbb{R}^J \to (-\infty, +\infty]$ be a convex function. Let h be another convex function, with effective domain D, that is differentiable on the nonempty open convex set int D. Assume that $f(x)$ is finite on $C = \overline{D}$ and attains its minimum value on C at $\hat{x}$. Our objective is to minimize $f(x)$ over x in $C = \overline{D}$.

At the kth step of a *proximal minimization algorithm* (PMA) [95, 55], we minimize the function

$$G_k(x) = f(x) + D_h(x, x^{k-1}), \tag{4.4}$$

to get x^k. The Bregman distance D_h is sometimes called a *proximity function*. The function

$$g_k(x) = D_h(x, x^{k-1}) \tag{4.5}$$

is nonnegative and $g_k(x^{k-1}) = 0$. We assume that each x^k lies in int D. As we shall see,

$$G_k(x) - G_k(x^k) \geq D_h(x, x^k) = g_{k+1}(x) \geq 0, \tag{4.6}$$

so the PMA is in the SUMMA class.

The Newton-Raphson algorithm for minimizing a twice differentiable function $f : \mathbb{R}^J \to \mathbb{R}$ has the iterative step

$$x^k = x^{k-1} - \nabla^2 f(x^{k-1})^{-1} \nabla f(x^{k-1}). \tag{4.7}$$

Suppose now that f is also convex. It is interesting to note that, having calculated x^{k-1}, we can obtain x^k by minimizing

$$G_k(x) = f(x) + (x - x^{k-1})^T \nabla^2 f(x^{k-1})(x - x^{k-1}) - D_f(x, x^{k-1}). \tag{4.8}$$

4.3 Some Obstacles

The PMA can present some computational obstacles. When we minimize $G_k(x)$ to get x^k we find that we must solve the equation

$$\nabla h(x^{k-1}) - \nabla h(x^k) \in \partial f(x^k), \tag{4.9}$$

where the set $\partial f(x)$ is the sub-differential of f at x. When $f(x)$ is differentiable, we must solve

$$\nabla f(x^k) + \nabla h(x^k) = \nabla h(x^{k-1}). \tag{4.10}$$

A modification of the PMA, called the IPA for *interior-point algorithm* [55, 63], is designed to overcome these computational obstacles. We discuss the IPA later in this chapter. Another modification of the PMA that is similar to the IPA is the *forward-backward splitting* (FBS) method to be discussed in a later chapter.

4.4 All PMA Are SUMMA

We show now that all PMA are in the SUMMA class. We remind the reader that $f(x)$ is now assumed to be convex.

Lemma 4.1 *For each k we have*

$$G_k(x) - G_k(x^k) \geq D_h(x, x^k) = g_{k+1}(x). \tag{4.11}$$

Proof: Since x^k minimizes $G_k(x)$ within the set D, we have

$$0 \in \partial f(x^k) + \nabla h(x^k) - \nabla h(x^{k-1}), \tag{4.12}$$

so that

$$\nabla h(x^{k-1}) = u^k + \nabla h(x^k), \tag{4.13}$$

for some u^k in $\partial f(x^k)$. Then

$$G_k(x) - G_k(x^k) = f(x) - f(x^k) + h(x) - h(x^k) - \langle \nabla h(x^{k-1}), x - x^k \rangle.$$

Now substitute, using Equation (4.13), to get

$$G_k(x) - G_k(x^k) = f(x) - f(x^k) - \langle u^k, x - x^k \rangle + D_h(x, x^k). \tag{4.14}$$

Therefore,

$$G_k(x) - G_k(x^k) \geq D_h(x, x^k),$$

since u^k is in $\partial f(x^k)$. ∎

4.5 Convergence of the PMA

From the discussion of the SUMMA we know that $\{f(x^k)\}$ is monotonically decreasing to $f(\hat{x})$. As we noted previously, if the sequence $\{x^k\}$ is bounded, and $\hat{x}$ is unique, we can conclude that $\{x^k\} \to \hat{x}$.

Suppose that $\hat{x}$ is not known to be unique, but can be chosen in D; this will be the case, of course, whenever D is closed. Then $G_k(\hat{x})$ is finite for each k. From the definition of $G_k(x)$ we have

$$G_k(\hat{x}) = f(\hat{x}) + D_h(\hat{x}, x^{k-1}). \tag{4.15}$$

From Equation (4.14) we have

$$G_k(\hat{x}) = G_k(x^k) + f(\hat{x}) - f(x^k) - \langle u^k, \hat{x} - x^k \rangle + D_h(\hat{x}, x^k). \tag{4.16}$$

Therefore,

$$D_h(\hat{x}, x^{k-1}) - D_h(\hat{x}, x^k) =$$

$$f(x^k) - f(\hat{x}) + D_h(x^k, x^{k-1}) + f(\hat{x}) - f(x^k) - \langle u^k, \hat{x} - x^k \rangle. \tag{4.17}$$

It follows that the sequence $\{D_h(\hat{x}, x^k)\}$ is decreasing and that $\{f(x^k)\}$ converges to $f(\hat{x})$. If either the function $f(x)$ or the function $D_h(\hat{x}, \cdot)$ has bounded level sets, then the sequence $\{x^k\}$ is bounded, has cluster points

x^* in C, and $f(x^*) = f(\hat{x})$, for every x^*. We now show that $\hat{x}$ in D implies that x^* is also in D, whenever h is a Bregman-Legendre function (see Appendix A).

Let x^* be an arbitrary cluster point, with $\{x^{k_n}\} \to x^*$. If $\hat{x}$ is not in the interior of D, then, by Property B2 of Bregman-Legendre functions, we know that

$$D_h(x^*, x^{k_n}) \to 0,$$

so x^* is in D. Then the sequence $\{D_h(x^*, x^k)\}$ is decreasing. Since a subsequence converges to zero, we have $\{D_h(x^*, x^k)\} \to 0$. From Property R5, we conclude that $\{x^k\} \to x^*$.

If $\hat{x}$ is in int D, but x^* is not, then $\{D_h(\hat{x}, x^k)\} \to +\infty$, by Property R2. But, this is a contradiction; therefore x^* is in D. Once again, we conclude that $\{x^k\} \to x^*$.

Now we summarize our results for the PMA. Let $f : \mathbb{R}^J \to (-\infty, +\infty]$ be closed, proper, and convex. Let h be a closed proper convex function, with effective domain D, that is differentiable on the nonempty open convex set int D. Assume that $f(x)$ is finite on $C = \overline{D}$ and attains its minimum value on C at $\hat{x}$. For each positive integer k, let x^k minimize the function $f(x) + D_h(x, x^{k-1})$. Assume that each x^k is in the interior of D.

Theorem 4.1 *If the restriction of $f(x)$ to x in C has bounded level sets and $\hat{x}$ is unique, then the sequence $\{x^k\}$ converges to $\hat{x}$.*

Theorem 4.2 *If $h(x)$ is a Bregman-Legendre function and $\hat{x}$ can be chosen in D, then $\{x^k\} \to x^*$, x^* in D, with $f(x^*) = f(\hat{x})$.*

4.6 The Non-differentiable Case

In the discussion so far, we have assumed that the function $h(x)$ is differentiable; the gradient played a role in the definition of the Bregman distance $D_h(x, z)$. When $h(x)$ is not differentiable, a PMA is still available. In the non-differentiable case a Bregman distance is defined to be

$$D_h(x, z; p) = h(x) - h(z) - \langle p, x - z \rangle , \tag{4.18}$$

where p is a member of the sub-differential $\partial h(z)$. We begin the PMA by selecting initial vectors x^0 and $p^0 \in \partial h(x^0)$. Now the iterate x^k minimizes

$$G_k(x) = f(x) + D_h(x, x^{k-1}; p^{k-1}), \tag{4.19}$$

where p^{k-1} is a member of $\partial h(x^{k-1})$. Therefore,

$$0 \in \partial f(x^k) + \partial h(x^k) - p^{k-1}. \tag{4.20}$$

We assume that this equation can be solved and that there are $u^k \in \partial f(x^k)$ and $v^k \in \partial h(x^k)$ so that

$$v^k = p^{k-1} - u^k. \tag{4.21}$$

We then define $p^k = v^k$, so that

$$G_k(x) - G_k(x^k) =$$

$$D_f(x, x^k; u^k) + D_h(x, x^k; p^k) \geq D_h(x, x^k; p^k) = g_{k+1}(x). \tag{4.22}$$

Therefore, the SUMMA condition holds and the sequence $\{f(x^k)\}$ converges to $f(\hat{x})$.

4.7 The IPA

The IPA is a modification of the PMA designed to overcome some of the computational obstacles encountered in the PMA [55, 63]. At the kth step of the PMA we must solve the equation

$$\nabla f(x^k) + \nabla h(x^k) = \nabla h(x^{k-1}) \tag{4.23}$$

for x^k, where, for notational convenience, we assume that both f and h are differentiable. Solving Equation (4.23) is probably not a simple matter, however. In the IPA approach we begin not with $h(x)$, but with a convex differentiable function $a(x)$ such that $h(x) = a(x) - f(x)$ is convex. Equation (4.23) now reads

$$\nabla a(x^k) = \nabla a(x^{k-1}) - \nabla f(x^{k-1}), \tag{4.24}$$

and we choose $a(x)$ so that Equation (4.24) is easily solved. We turn now to several examples of the IPA.

4.8 Projected Gradient Descent

The problem now is to minimize $f : \mathbb{R}^J \to \mathbb{R}$, over the closed, non-empty convex set C, where f is convex and differentiable on $\mathbb{R}^J$. We assume now that the gradient operator ∇f is L-Lipschitz continuous; that is, for all x and y, we have

$$\|\nabla f(x) - \nabla f(y)\|_2 \leq L\|x - y\|_2. \tag{4.25}$$

To employ the IPA approach, we let $0 < \gamma < \frac{1}{L}$ and select the function

$$a(x) = \frac{1}{2\gamma}\|x\|_2^2; \tag{4.26}$$

the upper bound on γ guarantees that the function $h(x) = a(x) - f(x)$ is convex. At the kth step we minimize

$$G_k(x) = f(x) + D_h(x, x^{k-1}) =$$

$$f(x) + \frac{1}{2\gamma}\|x - x^{k-1}\|_2^2 - D_f(x, x^{k-1}), \tag{4.27}$$

over $x \in C$. The solution x^k is in C and satisfies the inequality

$$\langle x^k - (x^{k-1} - \gamma\nabla f(x^{k-1})), c - x^k\rangle \geq 0, \tag{4.28}$$

for all $c \in C$. It follows then that

$$x^k = P_C(x^{k-1} - \gamma\nabla f(x^{k-1})); \tag{4.29}$$

here P_C denotes the orthogonal projection onto C. This is the projected gradient descent algorithm. For convergence we must require that f have certain additional properties needed for convergence of a PMA algorithm. Note that the auxiliary function

$$g_k(x) = \frac{1}{2\gamma}\|x - x^{k-1}\|_2^2 - D_f(x, x^{k-1}) \tag{4.30}$$

is unrelated to the set C, so is not used here to incorporate the constraint; it is used to provide a closed-form iterative scheme.

When $C = \mathbb{R}^J$ we have no constraint and the problem is simply to minimize f. Then the iterative algorithm becomes

$$x^k = x^{k-1} - \gamma\nabla f(x^{k-1}); \tag{4.31}$$

this is the gradient descent algorithm.

4.9 Relaxed Gradient Descent

In the gradient descent method we move away from the current x^{k-1} by the vector $\gamma\nabla f(x^{k-1})$. In relaxed gradient descent, the magnitude of the movement is reduced by α, where $\alpha \in (0, 1)$. Such relaxation methods are sometimes used to accelerate convergence. The relaxed gradient descent method can also be formulated as an AF method.

At the kth step we minimize

$$G_k(x) = f(x) + \frac{1}{2\gamma\alpha}\|x - x^{k-1}\|_2^2 - D_f(x, x^{k-1}), \tag{4.32}$$

obtaining

$$x^k = x^{k-1} - \alpha\gamma\nabla f(x^{k-1}). \tag{4.33}$$

4.10 Regularized Gradient Descent

In many applications the function to be minimized involves measured data, which is typically noisy, as well as some less than perfect model of how the measured data was obtained. In such cases, we may not want to minimize $f(x)$ exactly. In regularization methods we add to $f(x)$ another function that is designed to reduce sensitivity to noise and model error.

For example, suppose that we want to minimize

$$\alpha f(x) + \frac{1-\alpha}{2}\|x - p\|^2, \tag{4.34}$$

where p is chosen a priori. The regularized gradient descent algorithm for this problem can be put in the framework of a sequential unconstrained minimization problem.

At the kth step we minimize

$$G_k(x) = f(x) + \frac{1}{2\gamma\alpha}\|x - x^{k-1}\|_2^2 - \frac{1}{\alpha}(x, x^{k-1})] + \frac{1-\alpha}{2\gamma\alpha}\|x - p\|_2^2, \tag{4.35}$$

obtaining

$$x^k = \alpha(x^{k-1} - \gamma\nabla f(x^{k-1})) + (1-\alpha)p. \tag{4.36}$$

If we select $p = 0$ the iterative step becomes

$$x^k = \alpha(x^{k-1} - \gamma\nabla f(x^{k-1})). \tag{4.37}$$

4.11 The Projected Landweber Algorithm

The Landweber (LW) and projected Landweber (PLW) algorithms are special cases of projected gradient descent. The objective now is to minimize the function

$$f(x) = \frac{1}{2}\|Ax - b\|_2^2, \tag{4.38}$$

over $x \in \mathbb{R}^J$ or $x \in C$, where A is a real I by J matrix. The gradient of $f(x)$ is

$$\nabla f(x) = A^T(Ax - b) \tag{4.39}$$

and is L-Lipschitz continuous for $L = \rho(A^TA)$, the largest eiqenvalue of A^TA. The Bregman distance associated with $f(x)$ is

$$D_f(x, z) = \frac{1}{2}\|Ax - Az\|_2^2. \tag{4.40}$$

We let

$$a(x) = \frac{1}{2\gamma}\|x\|_2^2, \tag{4.41}$$

where $0 < \gamma < \frac{1}{L}$, so that the function $h(x) = a(x) - f(x)$ is convex.

At the kth step of the PLW we minimize

$$G_k(x) = f(x) + D_h(x, x^{k-1}) \tag{4.42}$$

over $x \in C$ to get

$$x^k = P_C(x^{k-1} - \gamma A^T(Ax^{k-1} - b)); \tag{4.43}$$

in the case of $C = \mathbb{R}^J$ we get the Landweber algorithm.

4.12 The Simultaneous MART

The simultaneous MART (SMART) minimizes the Kullback-Leibler distance $f(x) = KL(Px, y)$, where y is a positive vector, P is an I by J matrix with nonnegative entries P_{ij} for which $s_j = \sum_{i=1}^I P_{ij} = 1$, for all j, and we seek a nonnegative solution of the system $y = Px$.

The Bregman distance associated with the function $f(x) = KL(Px, y)$ is

$$D_f(x, z) = KL(Px, Pz). \tag{4.44}$$

We select $a(x)$ to be

$$a(x) = \sum_{j=1}^{J} x_j \log(x_j) - x_j. \tag{4.45}$$

It follows from the inequality in (2.13) that $h(x)$ is convex and

$$D_h(x, z) = KL(x, z) - KL(Px, Pz) \geq 0. \tag{4.46}$$

At the kth step of the SMART we minimize

$$G_k(x) = f(x) + D_h(x, x^{k-1}) =$$

$$KL(Px, y) + KL(x, x^{k-1}) - KL(Px, Px^{k-1}) \tag{4.47}$$

to get

$$x_j^k = x_j^{k-1} \exp\Big(\sum_{i=1}^{I} P_{ij} \log \frac{y_i}{(Px^{k-1})_i}\Big). \tag{4.48}$$

4.13 A Convergence Theorem

So far, we haven't discussed the restrictions necessary to prove convergence of these iterative algorithms. The IPA framework can be helpful in this regard, as we illustrate now.

The following theorem concerns convergence of the projected gradient descent algorithm with iterative step given by Equation (4.29).

Theorem 4.3 *Let $f : \mathbb{R}^J \to \mathbb{R}$ be differentiable, with L-Lipschitz continuous gradient. For γ in the interval $(0, \frac{1}{L})$ the sequence $\{x^k\}$ given by Equation (4.29) converges to a minimizer of f, over $x \in C$, whenever minimizers exist.*

Proof: The auxiliary function

$$g_k(x) = \frac{1}{2\gamma}\|x - x^{k-1}\|_2^2 - D_f(x, x^{k-1}) \tag{4.49}$$

can be rewritten as

$$g_k(x) = D_h(x, x^{k-1}), \tag{4.50}$$

where

$$h(x) = \frac{1}{2\gamma}\|x\|_2^2 - f(x). \tag{4.51}$$

Therefore, $g_k(x) \geq 0$ whenever $h(x)$ is a convex function.

We know that $h(x)$ is convex if and only if

$$\langle \nabla h(x) - \nabla h(y), x - y \rangle \geq 0, \tag{4.52}$$

for all x and y. This is equivalent to

$$\frac{1}{\gamma}\|x - y\|_2^2 - \langle \nabla f(x) - \nabla f(y), x - y \rangle \geq 0. \tag{4.53}$$

Since ∇f is L-Lipschitz, the inequality (4.53) holds whenever $0 < \gamma < \frac{1}{L}$.

A relatively simple calculation shows that

$$G_k(x) - G_k(x^k) =$$

$$\frac{1}{2\gamma}\|x - x^k\|_2^2 + \frac{1}{\gamma}\langle x^k - (x^{k-1} - \gamma\nabla f(x^{k-1})), x - x^k \rangle. \tag{4.54}$$

From Equation (4.29) it follows that

$$G_k(x) - G_k(x^k) \geq \frac{1}{2\gamma}\|x - x^k\|_2^2, \tag{4.55}$$

for all $x \in C$, so that, for all $x \in C$, we have

$$G_k(x) - G_k(x^k) \geq \frac{1}{2\gamma}\|x - x^k\|_2^2 - D_f(x, x^k) = g_{k+1}(x). \tag{4.56}$$

Now let $\hat{x}$ minimize $f(x)$ over all $x \in C$. Then

$$G_k(\hat{x}) - G_k(x^k) = f(\hat{x}) + g_k(\hat{x}) - f(x^k) - g_k(x^k)$$

$$\leq f(\hat{x}) + G_{k-1}(\hat{x}) - G_{k-1}(x^{k-1}) - f(x^k) - g_k(x^k),$$

so that

$$\Big(G_{k-1}(\hat{x}) - G_{k-1}(x^{k-1})\Big) - \Big(G_k(\hat{x}) - G_k(x^k)\Big) \geq f(x^k) - f(\hat{x}) + g_k(x^k) \geq 0.$$

Therefore, the sequence $\{G_k(\hat{x}) - G_k(x^k)\}$ is decreasing and the sequences $\{g_k(x^k)\}$ and $\{f(x^k) - f(\hat{x})\}$ converge to zero.

From

$$G_k(\hat{x}) - G_k(x^k) \geq \frac{1}{2\gamma}\|\hat{x} - x^k\|_2^2,$$

it follows that the sequence $\{x^k\}$ is bounded. Let $\{x^{k_n}\}$ converge to x^* with $\{x^{k_n+1}\}$ converging to x^{**}; we then have $f(x^*) = f(x^{**}) = f(\hat{x})$.

Replacing the generic $\hat{x}$ with x^{**}, we find that $\{G_{k_n+1}(x^{**}) - G_{k_n+1}(x^{k_n+1})\}$ is decreasing. By Equation (4.54), this subsequence converges to zero; therefore, the entire sequence $\{G_k(x^{**}) - G_k(x^k)\}$ converges to zero. From the inequality in (4.55), we conclude that the sequence $\{\|x^{**} - x^k\|_2^2\}$ converges to zero, and so $\{x^k\}$ converges to x^{**}. This completes the proof of the theorem. ∎

Using Theorem 6.3 it can be shown that convergence holds whenever γ is in the interval $(0, \frac{2}{L})$.

4.14 Another Job for the PMA

As we have seen, the original goal of the PMA is to minimize a convex function $f(x)$ over the closure of the domain of $h(x)$. Since the PMA is a SUMMA algorithm, we know that, whenever the sequence converges, the limit x^* satisfies $f(x^*) = d$, where d is the finite infimum of $f(x)$ over x in the interior of the domain of h. This suggests another job for the PMA.

Consider the problem of minimizing a differentiable convex function $h : \mathbb{R}^J \to \mathbb{R}$ over all x for which $Ax = b$, where A is an M by N matrix with rank M and b is arbitrary. With

$$f(x) = \frac{1}{2}\|Ax - b\|_2^2, \tag{4.57}$$

and x^0 arbitrary we minimize

$$f(x) + D_h(x, x^{k-1}) \tag{4.58}$$

to get x^k. Whenever the sequence $\{x^k\}$ converges to some x^* we have $Ax^* = b$. If $\nabla h(x^0)$ is in the range of A^T, then so is $\nabla h(x^*)$ and x^* minimizes $h(x)$ over all x with $Ax = b$.

4.15 The Goldstein-Osher Algorithm

In [132] the authors consider the problem of minimizing the function

$$\|\Phi(u)\|_1 + H(u), \tag{4.59}$$

where both $\|\Phi(u)\|_1$ and $H(u)$ are convex. They reformulate the problem as a constrained minimization problem as follows: minimize

$$\|d\|_1 + H(u), \tag{4.60}$$

subject to $\Phi(u) - d = 0$. They note that one penalty-function approach would be to minimize

$$\|d\|_1 + H(u) + \frac{k}{2}\|\Phi(u) - d\|_2^2, \tag{4.61}$$

to get (u^k, d^k). As $k \to +\infty$ the second term becomes increasingly important and the limit (u^*, d^*) of such a sequence would surely have $\Phi(u^*) - d^* = 0$. As k grows larger, however, the minimization problem becomes increasingly unstable. To avoid this, they propose an alternative algorithm, which we shall call here the Goldstein-Osher, or GO, algorithm.

The GO algorithm begins with an arbitrary choice of the vector b^{-1}. Having found u^{k-1}, d^{k-1}, and b^{k-1}, the next iterate (u^k, d^k) minimizes

$$\|d\|_1 + H(u) + \frac{1}{2}\|\Phi(u) - d - b^{k-1}\|_2^2. \tag{4.62}$$

The next b^k is

$$b^k = b^{k-1} - (\Phi(u^k) - d^k). \tag{4.63}$$

It is clear how the penalty-function approach in Equation (4.61) forces $\Phi(u^*) = d^*$, but it is not obvious how changing b^{k-1} to b^k would have the same effect. Essentially what happens is that, as $k \to +\infty$, an increasingly large vector is added to $\Phi(u) - d$ prior to taking the Euclidean norm. This has an effect similar to taking k increasingly large in the penalty-function approach.

The following theorem is essentially their Theorem 2.2.

Theorem 4.4 *If, for some k, we have $\Phi(u^k) = d^k$, then (u^k, d^k) minimizes $\|d\|_1 + H(u)$, subject to $\Phi(u) - d = 0$.*

Proof: Let $(\hat{u}, \hat{d})$ satisfy $\Phi(\hat{u}) - \hat{d} = 0$, and minimize $\|d\|_1 + H(u)$ over all (u, d) with $\Phi(u) - d = 0$. Then

$$H(u^k, d^k) + \frac{1}{2}\|\Phi(u^k) - d^k - b^{k-1}\|_2^2 \leq H(\hat{u}, \hat{d}) + \frac{1}{2}\|\Phi(\hat{u}) - \hat{d} - b^{k-1})\|_2^2,$$

so that $H(u^k, d^k) \leq H(\hat{u}, \hat{d})$. ∎

It is certainly restrictive to assume that $\Phi(u^k) - d^k = 0$ for finite k, although this can sometimes happen [184]. Clearly, a better result would assert that the sequence $\{(u^k, d^k)\}$ converges to some (u^*, d^*) with $\Phi(u^*) - d^* = 0$. Then (u^*, d^*) would solve their original problem.

In order to strengthen their conclusions regarding the GO algorithm, they attempt to relate the GO algorithm to the proximal minimization algorithm (PMA) [94, 95]. They claim that, for any sequence $\{(u^k, d^k)\}$ generated by the GO algorithm, there is a PMA that produces the same sequence. Such a PMA would employ a Bregman distance D_E derived from the convex function

$$E(u, d) = \|d\|_1 + H(u).$$

If this claim were true generally, then the theory of the PMA would say that, if $\{(u^k, d^k)\}$ converges to some (u^*, d^*), then $\Phi(u^*) - d^* = 0$.

On the face of it, this claim seems problematic. The reason is that the PMA does not directly involve the function $E(u, d)$, but only D_E, and D_E does not uniquely determine E. How would the PMA know to provide a constrained minimizer for E and not for one of the infinitely many alternatives that give the same D_E? The answer would have to lie with the selection of the starting point for the PMA; unfortunately, this aspect of the problem is ignored in [132].

As we shall see, their claim appears to be valid only when $\Phi(u)$ is affine linear. In addition, as we shall prove later, with certain requirements placed on the functions $\Phi(u)$ and $H(u)$, if the sequence (u^k, d^k) does converge to some (u^*, d^*), then the sequence $\{b^k\}$ converges and $\Phi(u^*) - d^* = 0$. Consequently, in such cases an equivalent PMA iteration seems to be unnecessary.

In the affine linear case, the problem is to minimize

$$E(u, d) = \|d\|_1 + H(u), \tag{4.64}$$

over all (u, d) with $Au - d = 0$, where A is a matrix. We know that a necessary and sufficient condition for (u^*, d^*) with $Au^* - d^* = 0$ to solve the minimization problem is for there to be a vector $v^* = (r^*, s^*)$ in the sub-differential $\partial E(u^*, d^*)$ such that r^* is in the range of A^T. Since, for each k, we have

$$v^k = (r^k, s^k) \in \partial E(u^k, d^k), \tag{4.65}$$

with

$$r^k + +A^T(Au^k - d^k - b^{k-1}) = 0, \tag{4.66}$$

it follows that, whenever the sequence $\{(u^k, d^k)\}$ converges to some (u^*, d^*) with $Au^* - d^* = 0$, the vector (u^*, d^*) solves the constrained minimization problem.

For the remainder of this section we shall adopt different notation and formulate the original problem as follows. Let $T : \mathbb{R}^J \to R^I$ be a (possibly) nonlinear operator, and $h : \mathbb{R}^J \to \mathbb{R}$ a convex function. Let $S = \{x | T(x) = 0\} \neq \emptyset$. Then we want to minimize the function $h(x)$ over $x \in S$. We shall obtain some new results concerning the GO algorithm, discuss the PMA, and investigate the claim in [132] that each sequence generated by the GO algorithm is also a PMA sequence.

Let $h : \mathbb{R}^J \to \mathbb{R}$ be a convex function, and $T : \mathbb{R}^J \to \mathbb{R}^I$ a (possibly) non-linear operator. Let $\hat{x} \in S$ minimize $h(x)$ over all x in S. For each $x \in \mathbb{R}^J$ let $T'(x) \in \mathbb{R}^{J \times I}$ be the Jacobian matrix for T at x. We assume that the operators T and T' are continuous, that for each x there is an I by J matrix $B(x)$ such that $B(x)T'(x) = I$, the identity matrix, and that the operator $B : \mathbb{R}^J \to \mathbb{R}^{I \times J}$ is continuous. For example, suppose that $T(x)$ is affine linear, so that there are matrix A and vector b with $T(x) = Ax - b$. Assuming that AA^T is invertible, we have $T'(x) = A^T$ and $B(x) = (AA^T)^{-1}A$.

Now the GO algorithm takes the following form For arbitrary vector b^{-1}, and having found x^{k-1} and b^{k-1}, take x^k to be the minimizer of the function

$$h(x) + \frac{1}{2}\|T(x) - b^{k-1}\|_2^2. \tag{4.67}$$

Then there is $v^k \in \partial h(x^k)$ such that

$$v^k = T'(x^k)(T(x^k) - b^{k-1}), \tag{4.68}$$

and

$$b^{k-1} = B(x^k)v^k + T(x^k). \tag{4.69}$$

If the sequences $\{x^k\}$ and $\{v^k\}$ converge, then so does the sequence $\{b^k\}$. Note that we have not yet said how the next b^k is to be defined.

If h is differentiable, then we have

$$\nabla h(x^k) = T'(x^k)(T(x^k) - b^{k-1}), \tag{4.70}$$

so that

$$b^{k-1} = B(x^k)\nabla h(x^k) + T(x^k). \tag{4.71}$$

We have the following theorem.

Theorem 4.5 *If h is continuously differentiable, and the sequence $\{x^k\}$ converges to some x^*, then the sequence $\{b^k\}$ converges to some b^*. If $T(x^*) = 0$, then x^* minimizes $h(x)$ over x in S.*

Proof: The first assertion follows from Equation (4.71) and continuity. We have

$$h(x^k) + \frac{1}{2}\|T(x^k) - b^{k-1}\|_2^2 \leq h(\hat{x}) + \frac{1}{2}\|T(\hat{x}) - b^{k-1}\|_2^2,$$

so that, by taking limits, we have

$$h(x^*) + \frac{1}{2}\|b^*\|_2^2 \leq h(\hat{x}) + \frac{1}{2}\|b^*\|_2^2.$$

∎

This theorem is similar to Theorem 2.2 of [132]; the latter does not require that h be differentiable.

In the Goldstein-Osher algorithm we have

$$b^k = b^{k-1} - T(x^k). \tag{4.72}$$

Now we can strengthen Theorem 4.5.

Theorem 4.6 *Let h be differentiable. Let b^k be defined as in Equation (4.72). If the sequence $\{x^k\}$ converges to some x^*, then $T(x^*) = 0$. Consequently, x^* minimizes $h(x)$ over x in S. If h is not differentiable, but the sequence $\{v^k\}$ converges, then the same result holds.*

Proof: We know that the sequence $\{b^k\}$ converges, according to Theorem 4.5 or the convergence of the sequence $\{v^k\}$. Therefore, by taking limits in Equation (4.72), we have $T(x^*) = 0$. ∎

In [132] the authors claim that, for each sequence generated by their algorithm, there is a proximal minimization algorithm that generates the same sequence. If true, this would allow them to make additional assertions about their algorithm. As we shall show, however, this equivalence appears to be true only when $T(x)$ is affine linear, that is, $T(x) = Ax - b$, so that $T'(x) = A^T$. For simplicity, we consider the case of differentiable h. Let b^{-1} be arbitrary. Let z^0 minimize the function

$$h(x) + \frac{1}{2}\|T(x) - b^{-1}\|_2^2, \tag{4.73}$$

and

$$b^0 = b^{-1} - T(z^0). \tag{4.74}$$

Then

$$\nabla h(z^0) + T'(z^0)(T(z^0) - b^{-1}) = \nabla h(z^0) - T'(z^0)b^0 = 0 \tag{4.75}$$

or

$$\nabla h(z^0) = T'(z^0)b^0. \tag{4.76}$$

Similarly,

$$\nabla h(z^1) + T'(z^1)(T(z^1) - b^0) = h(z^1) - T'(z^1)b^1 = 0, \tag{4.77}$$

so

$$\nabla h(z^1) = T'(z^1)b^1, \tag{4.78}$$

and

$$\nabla h(z^2) + T'(z^2)T(z^2) = T'(z^2)b^1. \tag{4.79}$$

Now consider the PMA.

Let x^1 minimize

$$\frac{1}{2}\|T(x)\|_2^2 + D_h(x, x^0), \tag{4.80}$$

for some x^0. Then

$$\nabla h(x^0) = \nabla h(x^1) + T'(x^1)T(x^1). \tag{4.81}$$

Similarly,

$$\nabla h(x^1) = \nabla h(x^2) + T'(x^2)T(x^2). \tag{4.82}$$

If $x^1 = z^1$ and $x^2 = z^2$ then

$$\nabla h(x^2) + T'(x^2)T(x^2) = T'(x^2)b^1, \tag{4.83}$$

or

$$\nabla h(x^1) = T'(x^2)b^1. \tag{4.84}$$

But we also have

$$\nabla h(x^1) = T'(x^1)b^1. \tag{4.85}$$

This suggests that there will be an equivalent PMA only when $T'(x)$ is constant, or $T(x) = Ax - b$; that is, only when $T(x)$ is affine linear.

4.16 A Question

Suppose that x^k minimizes

$$f(x) + D_h(x, x^{k-1}),$$

and $\{x^k\}$ converges to x^*. We know that x^* minimizes $f(x)$ over all x in the closure of the essential domain of h. Let M be the set of all such minimizers. Does x^* also minimize $h(z)$ over all z in M? In general, the answer is no; D_h does not determine h uniquely. What if $h(x) = D_h(x, x^0)$? There are several examples, using both Euclidean and Kullback-Leibler distances, in which the answer is yes.

Chapter 5

The Forward-Backward Splitting Algorithm

The *forward-backward splitting* (FBS) methods [101, 67] form a broad class of SUMMA algorithms closely related to the IPA. Note that minimizing $G_k(x)$ in Equation (4.4) over $x \in C$ is equivalent to minimizing

$$G_k(x) - \iota_C(x) + f(x) + D_h(x, x^{k-1}) \tag{5.1}$$

over all $x \in \mathbb{R}^J$, where $\iota_C(x) = 0$ for $x \in C$ and $\iota_C(x) = +\infty$ otherwise. This suggests a more general iterative algorithm, the FBS.

Suppose that we want to minimize the function $f_1(x) + f_2(x)$, where both functions are convex and $f_2(x)$ is differentiable with its gradient L-Lipschitz continuous in the Euclidean norm, by which we mean that

$$\|\nabla f_2(x) - \nabla f_2(y)\|_2 \leq L\|x - y\|_2, \tag{5.2}$$

for all x and y. At the kth step of the FBS algorithm we obtain x^k by minimizing

$$G_k(x) = f_1(x) + f_2(x) + \frac{1}{2\gamma}\|x - x^{k-1}\|_2^2 - D_{f_2}(x, x^{k-1}), \tag{5.3}$$

over all $x \in \mathbb{R}^J$, where $0 < \gamma < \frac{1}{2\gamma}$.

5.1 Moreau's Proximity Operators

Following Combettes and Wajs [101], we say that the *Moreau envelope* of index $\gamma > 0$ of the closed, proper convex function $f(x)$ is the continuous convex function

$$g(x) = \inf\{f(y) + \frac{1}{2\gamma}||x - y||_2^2\}, \tag{5.4}$$

with the infimum taken over all y in $\mathbb{R}^J$ [169, 170, 171]. In Rockafellar's book [192] and elsewhere, it is shown that the infimum is attained at a unique y, usually denoted $\text{prox}_{\gamma f}(x)$. The proximity operators $\text{prox}_{\gamma f}(\cdot)$ are firmly nonexpansive [101]; indeed, the proximity operator prox_f is the resolvent of the maximal monotone operator $B(x) = \partial f(x)$ and all such resolvent operators are firmly nonexpansive [35]. Proximity operators also generalize the orthogonal projections onto closed, convex sets: consider the function $f(x) = \iota_C(x)$, the *indicator function* of the closed, convex set C, taking the value zero for x in C, and $+\infty$ otherwise. Then $\text{prox}_{\gamma f}(x) = P_C(x)$, the orthogonal projection of x onto C. The following characterization of $x = \text{prox}_f(z)$ is quite useful: $x = \text{prox}_f(z)$ if and only if $z - x \in \partial f(x)$.

5.2 The FBS Algorithm

Our objective here is to provide an elementary proof of convergence for the forward-backward splitting (FBS) algorithm; a detailed discussion of this algorithm and its history is given by Combettes and Wajs in [101].

Let $f : \mathbb{R}^J \to \mathbb{R}$ be convex, with $f = f_1 + f_2$, both convex, f_2 differentiable, and ∇f_2 L-Lipschitz continuous. The iterative step of the FBS algorithm is

$$x^k = \text{prox}_{\gamma f_1}\Big(x^{k-1} - \gamma \nabla f_2(x^{k-1})\Big). \tag{5.5}$$

As we shall show, convergence of the sequence $\{x^k\}$ to a solution can be established, if γ is chosen to lie within the interval $(0, 1/L]$.

5.3 Convergence of the FBS Algorithm

Let $f : \mathbb{R}^J \to \mathbb{R}$ be convex, with $f = f_1 + f_2$, both convex, f_2 differentiable, and ∇f_2 L-Lipschitz continuous. Let $\{x^k\}$ be defined by Equation (5.5) and let $0 < \gamma \leq 1/L$.

For each $k = 1, 2, ...$ let

$$G_k(x) = f(x) + \frac{1}{2\gamma}\|x - x^{k-1}\|_2^2 - D_{f_2}(x, x^{k-1}), \tag{5.6}$$

where

$$D_{f_2}(x, x^{k-1}) = f_2(x) - f_2(x^{k-1}) - \langle \nabla f_2(x^{k-1}), x - x^{k-1} \rangle. \tag{5.7}$$

Since $f_2(x)$ is convex, $D_{f_2}(x, y) \geq 0$ for all x and y and is the Bregman distance formed from the function f_2.

The auxiliary function

$$g_k(x) = \frac{1}{2\gamma}\|x - x^{k-1}\|_2^2 - D_{f_2}(x, x^{k-1}) \tag{5.8}$$

can be rewritten as

$$g_k(x) = D_h(x, x^{k-1}), \tag{5.9}$$

where

$$h(x) = \frac{1}{2\gamma}\|x\|_2^2 - f_2(x). \tag{5.10}$$

Therefore, $g_k(x) \geq 0$ whenever $h(x)$ is a convex function.

We know that $h(x)$ is convex if and only if

$$\langle \nabla h(x) - \nabla h(y), x - y \rangle \geq 0, \tag{5.11}$$

for all x and y. This is equivalent to

$$\frac{1}{\gamma}\|x - y\|_2^2 - \langle \nabla f_2(x) - \nabla f_2(y), x - y \rangle \geq 0. \tag{5.12}$$

Since ∇f_2 is L-Lipschitz, the inequality (5.12) holds for $0 < \gamma \leq 1/L$.

Lemma 5.1 *The x^k that minimizes $G_k(x)$ over x is given by Equation (5.5).*

Proof: We know that x^k minimizes $G_k(x)$ if and only if

$$0 \in \nabla f_2(x^k) + \frac{1}{\gamma}(x^k - x^{k-1}) - \nabla f_2(x^k) + \nabla f_2(x^{k-1}) + \partial f_1(x^k),$$

or, equivalently,

$$\Big(x^{k-1} - \gamma \nabla f_2(x^{k-1})\Big) - x^k \in \partial(\gamma f_1)(x^k).$$

Consequently,

$$x^k = \text{prox}_{\gamma f_1}(x^{k-1} - \gamma \nabla f_2(x^{k-1})).$$

∎

Theorem 5.1 *The sequence $\{x^k\}$ converges to a minimizer of the function $f(x)$, whenever minimizers exist.*

Proof: A relatively simple calculation shows that

$$G_k(x) - G_k(x^k) = \frac{1}{2\gamma}\|x - x^k\|_2^2 +$$

$$\Big(f_1(x) - f_1(x^k) - \frac{1}{\gamma}\langle(x^{k-1} - \gamma\nabla f_2(x^{k-1})) - x^k, x - x^k\rangle\Big). \tag{5.13}$$

Since

$$(x^{k-1} - \gamma\nabla f_2(x^{k-1})) - x^k \in \partial(\gamma f_1)(x^k),$$

it follows that

$$\Big(f_1(x) - f_1(x^k) - \frac{1}{\gamma}\langle(x^{k-1} - \gamma\nabla f_2(x^{k-1})) - x^k, x - x^k\rangle\Big) \geq 0.$$

Therefore,

$$G_k(x) - G_k(x^k) \geq \frac{1}{2\gamma}\|x - x^k\|_2^2 \geq g_{k+1}(x). \tag{5.14}$$

Therefore, the inequality in (2.17) holds and the iteration fits into the SUMMA class.

Now let $\hat{x}$ minimize $f(x)$ over all x. Then

$$G_k(\hat{x}) - G_k(x^k) = f(\hat{x}) + g_k(\hat{x}) - f(x^k) - g_k(x^k)$$

$$\leq f(\hat{x}) + G_{k-1}(\hat{x}) - G_{k-1}(x^{k-1}) - f(x^k) - g_k(x^k),$$

so that

$$\Big(G_{k-1}(\hat{x}) - G_{k-1}(x^{k-1})\Big) - \Big(G_k(\hat{x}) - G_k(x^k)\Big) \geq f(x^k) - f(\hat{x}) + g_k(x^k) \geq 0.$$

Therefore, the sequence $\{G_k(\hat{x}) - G_k(x^k)\}$ is decreasing and the sequences $\{g_k(x^k)\}$ and $\{f(x^k) - f(\hat{x})\}$ converge to zero.

From

$$G_k(\hat{x}) - G_k(x^k) \geq \frac{1}{2\gamma}\|\hat{x} - x^k\|_2^2,$$

it follows that the sequence $\{x^k\}$ is bounded. Therefore, we may select a subsequence $\{x^{k_n}\}$ converging to some x^{**}, with $\{x^{k_n-1}\}$ converging to some x^*, and therefore $f(x^*) = f(x^{**}) = f(\hat{x})$.

Replacing the generic $\hat{x}$ with x^{**}, we find that $\{G_k(x^{**}) - G_k(x^k)\}$ is decreasing to zero. From the inequality in (5.14), we conclude that the sequence $\{\|x^* - x^k\|_2^2\}$ converges to zero, and so $\{x^k\}$ converges to x^*. This completes the proof of the theorem. ∎

5.4 Some Examples

We present some examples to illustrate the application of the convergence theorem.

5.4.1 Projected Gradient Descent

Let C be a nonempty, closed convex subset of $\mathbb{R}^J$ and $f_1(x) = \iota_C(x)$, the function that is $+\infty$ for x not in C and zero for x in C. Then $\iota_C(x)$ is convex, but not differentiable. We have $\text{prox}_{\gamma f_1} = P_C$, the orthogonal projection onto C. The iteration in Equation (5.5) becomes

$$x^k = P_C\Big(x^{k-1} - \gamma\nabla f_2(x^{k-1})\Big). \tag{5.15}$$

The sequence $\{x^k\}$ converges to a minimizer of f_2 over $x \in C$, whenever such minimizers exist, for $0 < \gamma \leq 1/L$.

5.4.2 The CQ Algorithm

Let A be a real I by J matrix, $C \subseteq \mathbb{R}^J$, and $Q \subseteq \mathbb{R}^I$, both closed convex sets. The split feasibility problem (SFP) is to find x in C such that Ax is in Q. The function

$$f_2(x) = \frac{1}{2}\|P_Q Ax - Ax\|_2^2 \tag{5.16}$$

is convex, differentiable and ∇f_2 is L-Lipschitz for $L = \rho(A^TA)$, the spectral radius of A^TA. The gradient of f_2 is

$$\nabla f_2(x) = A^T(I - P_Q)Ax. \tag{5.17}$$

We want to minimize the function $f_2(x)$ over x in C, or, equivalently, to minimize the function $f(x) = \iota_C(x)+f_2(x)$. The projected gradient descent algorithm has the iterative step

$$x^k = P_C\Big(x^{k-1} - \gamma A^T(I - P_Q)Ax^{k-1}\Big); \tag{5.18}$$

this iterative method was called the CQ-algorithm in [58, 59]. The sequence $\{x^k\}$ converges to a solution whenever f_2 has a minimum on the set C, for $0 < \gamma \leq 1/L$.

In [85, 81] the CQ algorithm was extended to a multiple-sets algorithm and applied to the design of protocols for intensity-modulated radiation therapy.

5.4.3 The Projected Landweber Algorithm

The problem is to minimize the function

$$f_2(x) = \frac{1}{2}\|Ax - b\|_2^2,$$

over $x \in C$. This is a special case of the SFP and we can use the CQ-algorithm, with $Q = \{b\}$. The resulting iteration is the projected Landweber algorithm [23]; when $C = \mathbb{R}^J$ it becomes the Landweber algorithm [154].

5.5 Minimizing f_2 over a Linear Manifold

Suppose that we want to minimize f_2 over x in the linear manifold $M = S + p$, where S is a subspace of $\mathbb{R}^J$ of dimension $I < J$ and p is a fixed vector. Let A be an I by J matrix such that the I columns of A^T form a basis for S. For each $z \in \mathbb{R}^I$ let

$$d(z) = f_2(A^T z + p),$$

so that d is convex, differentiable, and its gradient,

$$\nabla d(z) = A\nabla f_2(A^T z + p),$$

is K-Lipschitz continuous, for $K = \rho(A^T A)L$. The sequence $\{z^k\}$ defined by

$$z^k = z^{k-1} - \gamma\nabla d(z^{k-1}) \tag{5.19}$$

converges to a minimizer of d over all z in $\mathbb{R}^I$, whenever minimizers exist, for $0 < \gamma \leq 1/K$.

From Equation (5.19) we get

$$x^k = x^{k-1} - \gamma A^T A \nabla f_2(x^{k-1}), \tag{5.20}$$

with $x^k = A^T z^k + p$. The sequence $\{x^k\}$ converges to a minimizer of f_2 over all x in M.

Suppose now that we begin with an algorithm having the iterative step

$$x^k = x^{k-1} - \gamma A^T A \nabla f_2(x^{k-1}), \tag{5.21}$$

where A is any real I by J matrix having rank I. Let x^0 be in the range of A^T, so that $x^0 = A^T z^0$, for some $z^0 \in \mathbb{R}^I$. Then each $x^k = A^T z^k$ is again in the range of A^T, and we have

$$A^T z^k = A^T z^{k-1} - \gamma A^T A \nabla f_2(A^T z^{k-1}). \tag{5.22}$$

With $d(z) = f_2(A^T z)$, we can write Equation (5.22) as

$$A^T \Big(z^k - (z^{k-1} - \gamma \nabla d(z^{k-1})) \Big) = 0. \tag{5.23}$$

Since A has rank I, A^T is one-to-one, so that

$$z^k - z^{k-1} - \gamma \nabla d(z^{k-1}) = 0. \tag{5.24}$$

The sequence $\{z^k\}$ converges to a minimizer of d, over all $z \in \mathbb{R}^I$, whenever such minimizers exist, for $0 < \gamma \leq 1/K$. Therefore, the sequence $\{x^k\}$ converges to a minimizer of f_2 over all x in the range of A^T.

5.6 Feasible-Point Algorithms

Suppose that we want to minimize a convex differentiable function $f(x)$ over x such that $Ax = b$, where A is an I by J full-rank matrix, with $I < J$. If $Ax^k = b$ for each of the vectors $\{x^k\}$ generated by the iterative algorithm, we say that the algorithm is a feasible-point method.

5.6.1 The Projected Gradient Algorithm

Let C be the feasible set of all x in $\mathbb{R}^J$ such that $Ax = b$. For every z in $\mathbb{R}^J$, we have

$$P_C z = P_{NS(A)} z + A^T (AA^T)^{-1} b, \tag{5.25}$$

where $NS(A)$ is the null space of A. Using

$$P_{NS(A)}z = z - A^T(AA^T)^{-1}Az, \tag{5.26}$$

we have

$$P_C z = z + A^T(AA^T)^{-1}(b - Az). \tag{5.27}$$

Using Equation (5.5), we get the iteration step for the projected gradient algorithm:

$$x^k = x^{k-1} - \gamma P_C \nabla f(x^{k-1}), \tag{5.28}$$

which converges to a solution for $0 < \gamma \leq 1/L$, whenever solutions exist.

Next we present a somewhat simpler approach.

5.6.2 The Reduced Gradient Algorithm

Let x^0 be a feasible point, that is, $Ax^0 = b$. Then $x = x^0 + p$ is also feasible if p is in the null space of A, that is, $Ap = 0$. Let Z be a J by $J - I$ matrix whose columns form a basis for the null space of A. We want $p = Zv$ for some v. The best v will be the one for which the function

$$\phi(v) = f(x^0 + Zv)$$

is minimized. We can apply to the function $\phi(v)$ the steepest descent method, or the Newton-Raphson method, or any other minimization technique.

The steepest descent method, applied to $\phi(v)$, is called the reduced steepest descent algorithm [174]. The gradient of $\phi(v)$, also called the reduced gradient, is

$$\nabla\phi(v) = Z^T \nabla f(x),$$

where $x = x^0 + Zv$; the gradient operator $\nabla\phi$ is then K-Lipschitz, for $K = \rho(A^T A)L$.

Let x^0 be feasible. The iteration in Equation (5.5) now becomes

$$v^k = v^{k-1} - \gamma\nabla\phi(v^{k-1}), \tag{5.29}$$

so that the iteration for $x^k = x^0 + Zv^k$ is

$$x^k = x^{k-1} - \gamma ZZ^T \nabla f(x^{k-1}). \tag{5.30}$$

The vectors x^k are feasible and the sequence $\{x^k\}$ converges to a solution, whenever solutions exist, for any $0 < \gamma < \frac{1}{K}$.

5.6.3 The Reduced Newton-Raphson Method

The same idea can be applied to the Newton-Raphson method. The Newton-Raphson method, applied to $\phi(v)$, is called the reduced Newton-Raphson method [174]. The Hessian matrix of $\phi(v)$, also called the reduced Hessian matrix, is

$$\nabla^2\phi(v) = Z^T\nabla^2 f(c)Z,$$

so that the reduced Newton-Raphson iteration becomes

$$x^k = x^{k-1} - Z\Big(Z^T\nabla^2 f(x^{k-1})Z\Big)^{-1} Z^T\nabla f(x^{k-1}). \tag{5.31}$$

Let c^0 be feasible. Then each x^k is feasible. The sequence $\{x^k\}$ is not guaranteed to converge.

Chapter 6

Operators

6.1 Overview

In a broad sense, all iterative algorithms generate a sequence $\{x^k\}$ of vectors that describe the current state of the iterative process. The sequence may converge for any starting vector x^0, or may converge only if the x^0 is sufficiently close to a solution. The limit, when it exists, may depend on x^0, and may, or may not, solve the original problem. Convergence to the limit may be slow and the algorithm may need to be accelerated. The algorithm may involve measured data. The limit may be sensitive to noise in the data and the algorithm may need to be regularized to lessen this sensitivity. The algorithm may be quite general, applying to all problems in a broad class, or it may be tailored to the problem at hand. Each step of the algorithm may be costly, but only a few steps generally are needed to produce a suitable approximate answer, or, each step may be easily performed, but many such steps are needed. Although convergence of an algorithm is important, theoretically, sometimes in practice only a few iterative steps are used. In this chapter we consider several classes of operators that play important roles in optimization.

6.2 Operators

A function $T : \mathbb{R}^J \to \mathbb{R}^J$ is called an *operator* on $\mathbb{R}^J$. For our purposes, the most important examples of operators on $\mathbb{R}^J$ are the orthogonal projections P_C onto closed convex sets, and gradient operators, that is, $T(x) = \nabla g(x)$, for some differentiable function $g(x) : \mathbb{R}^J \to \mathbb{R}$. As we shall see later, the operators P_C are also gradient operators.

For many of the iterative algorithms we consider in this book, the iterative step is

$$x^{k+1} = Tx^k, \tag{6.1}$$

for some fixed operator T. If T is a continuous operator (and it usually is), and the sequence $\{T^k x^0\}$ converges to $\hat{x}$, then $T\hat{x} = \hat{x}$, that is, $\hat{x}$ is a *fixed point* of the operator T. We denote by $\text{Fix}(T)$ the set of fixed points of T. The convergence of the iterative sequence $\{T^k x^0\}$ will depend on the properties of the operator T.

Our approach here will be to identify several classes of operators for which the iterative sequence is known to converge, to examine the convergence theorems that apply to each class, to describe several applied problems that can be solved by iterative means, to present iterative algorithms for solving these problems, and to establish that the operator involved in each of these algorithms is a member of one of the designated classes.

6.3 Contraction Operators

Contraction operators are perhaps the best known class of operators associated with iterative algorithms.

6.3.1 Lipschitz Continuous Operators

Definition 6.1 *An operator T on $\mathbb{R}^J$ is* Lipschitz continuous, *with respect to a vector norm $||\cdot||$, or L-Lipschitz, if there is a positive constant L such that*

$$||Tx - Ty|| \leq L||x - y||, \tag{6.2}$$

for all x and y in $\mathbb{R}^J$.

For example, if $f : \mathbb{R} \to \mathbb{R}$, and $g(x) = f'(x)$ is differentiable, the Mean Value Theorem tells us that

$$g(b) = g(a) + g'(c)(b - a),$$

for some c between a and b. Therefore,

$$|f'(b) - f'(a)| \leq |f''(c)||b - a|.$$

If $|f''(x)| \leq L$, for all x, then $g(x) = f'(x)$ is L-Lipschitz. More generally, if $f : \mathbb{R}^J \to \mathbb{R}$ is twice differentiable and $\|\nabla^2 f(x)\|_2 \leq L$, for all x, then $T = \nabla f$ is L-Lipschitz, with respect to the 2-norm. The 2-norm of the Hessian matrix $\nabla^2 f(x)$ is the largest of the absolute values of its eigenvalues.

6.3.2 Nonexpansive Operators

An important special class of Lipschitz continuous operators are the nonexpansive, or contractive, operators.

Definition 6.2 *If $L = 1$, then T is said to be* nonexpansive *(ne), or a* contraction, *with respect to the given norm. In other words, T is ne for a given norm if, for every x and y, we have*

$$\|Tx - Ty\| \leq \|x - y\|.$$

Lemma 6.1 *Let $T : \mathbb{R}^J \to \mathbb{R}^J$ be a nonexpansive operator, with respect to the 2-norm. Then the set F of fixed points of T is a convex set.*

Proof: Select two distinct points a and b in F, a scalar α in the open interval $(0, 1)$, and let $c = \alpha a + (1 - \alpha)b$. We show that $Tc = c$. Note that

$$a - c = \frac{1 - \alpha}{\alpha}(c - b).$$

We have

$$\|a - b\| = \|a - Tc + Tc - b\| \leq \|a - Tc\| + \|Tc - b\| = \|Ta - Tc\| + \|Tc - Tb\|$$

$$\leq \|a - c\| + \|c - b\| = \|a - b\|;$$

the last equality follows since $a - c$ is a multiple of $(c - b)$. From this, we conclude that

$$\|a - Tc\| = \|a - c\|,$$

$$\|Tc - b\| = \|c - b\|,$$

and that $a - Tc$ and $Tc - b$ are positive multiples of one another, that is, there is $\beta > 0$ such that

$$a - Tc = \beta(Tc - b),$$

or

$$Tc = \frac{1}{1+\beta}a + \frac{\beta}{1+\beta}b = \gamma a + (1-\gamma)b.$$

Then inserting $c = \alpha a + (1-\alpha)b$ and $Tc = \gamma a + (1-\gamma)b$ into

$$\|Tc - b\| = \|c - b\|,$$

we find that $\gamma = \alpha$ and so $Tc = c$. ∎

The reader should note that the proof of the previous lemma depends heavily on the fact that the norm is the two-norm. If x and y are any nonnegative vectors then $\|x + y\|_1 = \|x\|_1 + \|y\|_1$, so the proof would not hold, if, for example, we used the one-norm instead.

We want to find properties of an operator T that guarantee that the sequence of iterates $\{T^k x_0\}$ will converge to a fixed point of T, for any x^0, whenever fixed points exist. Being nonexpansive is not enough; the nonexpansive operator $T = -I$, where $Ix = x$ is the identity operator, has the fixed point $x = 0$, but the sequence $\{T^k x^0\}$ converges only if $x^0 = 0$.

6.3.3 Strict Contractions

One property that guarantees not only that the iterates converge, but that there is a fixed point, is the property of being a strict contraction.

Definition 6.3 *An operator T on $\mathbb{R}^J$ is a* strict contraction *(sc), with respect to a vector norm $||\cdot||$, if there is $r \in (0,1)$ such that*

$$||Tx - Ty|| \leq r||x - y||, \tag{6.3}$$

for all vectors x and y.

For strict contractions, we have the Banach-Picard Theorem [114].

6.3.3.1 The Banach-Picard Theorem

Theorem 6.1 *Let T be sc. Then, there is a unique fixed point of T and, for any starting vector x^0, the sequence $\{T^k x^0\}$ converges to the fixed point.*

The key step in the proof is to show that $\{x^k\}$ is a Cauchy sequence, therefore, it has a limit.

Corollary 6.1 *If T^n is a strict contraction, for some positive integer n, then T has a fixed point.*

Proof: The proof is left as Exercise 6.14. ∎

In many of the applications of interest to us, there will be multiple fixed points of T. Therefore, T will not be sc for any vector norm, and the Banach-Picard fixed-point theorem will not apply. We need to consider other classes of operators. These classes of operators will emerge as we investigate the properties of orthogonal projection operators.

6.3.4 Instability

Suppose we rewrite the equation $e^{-x} = x$ as $x = -\log x$, and define $Tx = -\log x$, for $x > 0$. Now our iterative scheme becomes $x_{k+1} = Tx_k = -\log x_k$. A few calculations will convince us that the sequence $\{x_k\}$ is diverging away from the correct answer, not converging to it. The lesson here is that we cannot casually reformulate our problem as a fixed-point problem and expect the iterates to converge to the answer. What matters is the behavior of the operator T.

6.4 Convex Sets in $\mathbb{R}^J$

We begin with the basic definitions.

Definition 6.4 *A vector z is said to be a* convex combination *of the vectors x and y if there is α in the interval $[0,1]$ such that $z = (1-\alpha)x + \alpha y$. More generally, a vector z is a convex combination of the vectors x^n, $n = 1, ..., N$, if there are numbers $\alpha_n \geq 0$ with*

$$\alpha_1 + ... + \alpha_N = 1$$

and

$$z = \alpha_1 x^1 + ... + \alpha_N x^N.$$

Definition 6.5 *A nonempty set C in $\mathbb{R}^J$ is said to be* convex *if, for any distinct points x and y in C, and for any real number α in the interval $(0,1)$, the point $(1-\alpha)x + \alpha y$ is also in C; that is, C is closed to convex combinations of any two members of C.*

In Exercise 6.1 the reader is asked to show that if C is convex then the convex combination of any number of members of C is again in C. We say then that C is *closed to convex combinations.*

For example, the two-norm unit ball B in $\mathbb{R}^J$, consisting of all x with $||x||_2 \leq 1$, is convex, while the surface of the ball, the set of all x with $||x||_2 = 1$, is not convex. More generally, the unit ball of $\mathbb{R}^J$ in any norm is a convex set, as a consequence of the triangle inequality for norms.

Definition 6.6 *The* convex hull *of a set S, denoted conv(S), is the smallest convex set containing S, by which we mean that if K is any convex set containing S, then K must also contain conv(S).*

One weakness of this definition is that it does not tell us explicitly what the members of conv(S) look like, nor precisely how the individual members of conv(S) are related to the members of S itself. In fact, it is not obvious that a smallest such set exists at all. The following proposition remedies this; the reader is asked to supply a proof in Exercise 6.2 later.

Proposition 6.1 *The convex hull of a set S is the set C of all convex combinations of members of S.*

Definition 6.7 *A subset S of $\mathbb{R}^J$ is a* subspace *if, for every x and y in S and scalars α and β, the linear combination $\alpha x + \beta y$ is again in S.*

A subspace is necessarily a convex set.

Definition 6.8 *Let C be a nonempty convex subset of $\mathbb{C}^J$. A point x in C is said to be an* extreme point *of C if*

$$x = \alpha y + (1 - \alpha)z,$$

for some α in the interval $(0, 1)$ and y and z in C implies that $y = z = x$.

Said differently, x is an extreme point of C if x is not a member of any line segment between two other members of C.

A function $f : \mathbb{R}^J \to (-\infty, \infty]$ is convex if and only if its epigraph is a convex set in $\mathbb{R}^{J+1}$. At the same time, every closed convex set $C \subseteq \mathbb{R}^J$ has the form

$$C = \{x | f(x) \leq 0\}, \tag{6.4}$$

for some convex function $f : \mathbb{R}^J \to \mathbb{R}$. We are tempted to assume that the smoothness of the function f will be reflected in the geometry of the set C. In particular, we may well expect that, if x is on the boundary of C and f is differentiable at x, then there is a unique hyperplane supporting C at x and $\nabla f(x)$ is a non-zero normal vector; but this is wrong. Any closed convex nonempty set C can be written as in Equation (6.4), for the differentiable function

$$f(x) = \frac{1}{2}\|x - P_C x\|^2.$$

As we shall see later, the gradient of $f(x)$ is $\nabla f(x) = x - P_C x$, so that $\nabla f(x) = 0$ for every x in C. Nevertheless, the set C may have a unique supporting hyperplane at each boundary point, or it may have multiple such hyperplanes, regardless of the properties of the f used to define C.

When we first encounter gradients, usually in Calculus III, they are almost always described geometrically as a vector that is a normal for the hyperplane that is tangent to the level surface of f at that point, and as indicating the direction of greatest increase of f. However, this is not always the case.

Consider the function $f : \mathbb{R}^2 \to \mathbb{R}$ given by

$$f(x_1, x_2) = \frac{1}{2}(\sqrt{x_1^2 + x_2^2} - 1)^2,$$

for $x_1^2 + x_2^2 \geq 1$, and zero, otherwise. This function is differentiable and

$$\nabla f(x) = \frac{\|x\|_2 - 1}{\|x\|_2} x,$$

for $\|x\|_2 \geq 1$, and $\nabla f(x) = 0$, otherwise. The level surface in $\mathbb{R}^2$ of all x such that $f(x) \leq 0$ is the closed unit ball; it is not a simple closed curve. At every point of its boundary the gradient is zero, and yet, at each boundary point, there is a unique supporting tangent line.

Consider the function $f : \mathbb{R}^2 \to \mathbb{R}$ given by $f(x) = f(x_1, x_2) = x_1^2$. The level curve $C = \{x | f(x) = 0\}$ is the x_2 axis. For any x such that $x_1 = 0$ the hyperplane supporting C at x is C itself, and any vector of the form $(\gamma, 0)$ is a normal to C. But the gradient of $f(x)$ is zero at all points of C. So the gradient of f is not a normal vector to the supporting hyperplane.

6.5 Orthogonal Projection Operators

The following proposition is fundamental in the study of convexity and can be found in most books on the subject; see, for example, the text by Goebel and Reich [131].

Proposition 6.2 *Given any nonempty closed convex set C and an arbitrary vector x in $\mathbb{R}^J$, there is a unique member $P_C x$ of C closest, in the sense of the two-norm, to x. The vector $P_C x$ is called the* orthogonal (or metric) projection *of x onto C and the operator P_C the* orthogonal projection *onto C.*

Proof: If x is in C, then $P_C x = x$, so assume that x is not in C. Then $d > 0$, where d is the distance from x to C. For each positive integer n, select c^n in C with $\|x - c^n\|_2 < d + \frac{1}{n}$. Then, since for all n we have

$$\|c^n\|_2 = \|c^n - x + x\|_2 \leq \|c^n - x\|_2 + \|x\|_2 \leq d + \frac{1}{n} + \|x\|_2 < d + 1 + \|x\|_2,$$

the sequence $\{c^n\}$ is bounded; let c^* be any cluster point. It follows easily that $||x - c^*||_2 = d$ and that c^* is in C. If there is any other member c of C with $||x - c||_2 = d$, then, by the Parallelogram Law, we would have $||x - (c^* + c)/2||_2 < d$, which is a contradiction. Therefore, c^* is $P_C x$. ∎

The proof just given relies on the Bolzano-Weierstrass Theorem. There is another proof, which avoids this theorem and so is valid for infinite-dimensional Hilbert space. The idea is to use the Parallelogram Law to show that the sequence $\{c^n\}$ is Cauchy and then to use completeness to get c^*. We leave the details to the reader.

If C is a subspace, then we can get an explicit description of $P_C x$ in terms of x; for general convex sets C, however, we will not be able to express $P_C x$ explicitly, and certain approximations will be needed. Orthogonal projection operators are central to our discussion, and, in this overview, we focus on problems involving convex sets, algorithms involving orthogonal projection onto convex sets, and classes of operators derived from properties of orthogonal projection operators.

For an arbitrary nonempty closed convex set C in $\mathbb{R}^J$, the orthogonal projection $T = P_C$ is a nonlinear operator, unless, of course, C is a subspace. We may not be able to describe $P_C x$ explicitly, but we do know a useful property of $P_C x$.

Proposition 6.3 *For a given x, a vector z in C is $P_C x$ if and only if*

$$\langle c - z, z - x \rangle \geq 0, \tag{6.5}$$

for all c in the set C.

Proof: Let c be arbitrary in C and α in $(0, 1)$. Then

$$||x - P_C x||_2^2 \leq ||x - (1 - \alpha)P_C x - \alpha c||_2^2 = ||x - P_C x + \alpha(P_C x - c)||_2^2$$

$$= ||x - P_C x||_2^2 - 2\alpha\langle x - P_C x, c - P_C x\rangle + \alpha^2||P_C x - c||_2^2. \tag{6.6}$$

Therefore,

$$-2\alpha\langle x - P_C x, c - P_C x\rangle + \alpha^2||P_C x - c||_2^2 \geq 0, \tag{6.7}$$

so that

$$2\langle x - P_C x, c - P_C x\rangle \leq \alpha||P_C x - c||_2^2. \tag{6.8}$$

Taking the limit, as $\alpha \to 0$, we conclude that

$$\langle c - P_C x, P_C x - x\rangle \geq 0. \tag{6.9}$$

If z is a member of C that also has the property

$$\langle c - z, z - x\rangle \geq 0, \tag{6.10}$$

for all c in C, then we have both

$$\langle z - P_C x, P_C x - x \rangle \geq 0, \tag{6.11}$$

and

$$\langle z - P_C x, x - z \rangle \geq 0. \tag{6.12}$$

Adding on both sides of these two inequalities leads to

$$\langle z - P_C x, P_C x - z \rangle \geq 0. \tag{6.13}$$

But,

$$\langle z - P_C x, P_C x - z \rangle = -||z - P_C x||_2^2, \tag{6.14}$$

so it must be the case that $z = P_C x$. This completes the proof. ∎

Corollary 6.2 *For any x and y in $\mathbb{R}^J$ we have*

$$\langle P_C x - P_C y, x - y \rangle \geq \|P_C x - P_C y\|_2^2. \tag{6.15}$$

Proof: The proof is left as Exercise 6.15. ∎

The operators P_C have the property that

$$\langle x - P_C x, x \rangle \geq \langle x - P_C x, P_C x \rangle \geq \langle x - P_C x, c \rangle, \tag{6.16}$$

for all $c \in C$. This property suggests the following definition:

Definition 6.9 *An operator T on $\mathbb{C}^J$ is called a* cutter *operator if T has a fixed point and*

$$\langle x - Tx, Tx \rangle \geq \langle x - Tx, z \rangle, \tag{6.17}$$

for all fixed points z of T.

The term *cutter* was used in [78]; such operators were called *separating operators* in [76] and have had other names in the literature (see [77]).

Definition 6.10 *An operator U on $\mathbb{C}^J$ is* quasi-nonexpansive *(qne), with respect to some norm, if U has a fixed point and if*

$$\|x - z\| \geq \|Ux - z\|, \tag{6.18}$$

for all x and for all fixed points z of U.

Proposition 6.4 *(Corollary 2.1.33 of [77]) If an operator U on $\mathbb{C}^J$ has fixed points, then U is qne, with respect to the two-norm, if and only if $T = \frac{1}{2}(I + U)$ is a cutter operator.*

In Exercise 6.16 the reader is asked to prove Proposition 6.4.

It follows from Corollary 6.2 and Cauchy's Inequality that the orthogonal projection operator $T = P_C$ is nonexpansive, with respect to the Euclidean norm, that is,

$$||P_Cx - P_Cy||_2 \leq ||x - y||_2, \tag{6.19}$$

for all x and y. Because the operator P_C has multiple fixed points, P_C cannot be a strict contraction, unless the set C is a singleton set.

Corollary 6.2 tells us that the operators P_C are more than simply nonexpansive; they are *firmly nonexpansive.* A good source for more material on these topics are the books by Goebel and Reich [131] and by Bauschke and Combettes [18].

Definition 6.11 *An operator T is said to be* firmly nonexpansive *(fne) if*

$$\langle Tx - Ty, x - y\rangle \geq ||Tx - Ty||_2^2, \tag{6.20}$$

for all x and y in $\mathbb{R}^J$.

Lemma 6.2 *An operator $F : \mathbb{R}^J \to \mathbb{R}^J$ is fne if and only if $F = \frac{1}{2}(I + N)$, for some operator N that is ne with respect to the two-norm.*

Proof: Suppose that $F = \frac{1}{2}(I + N)$. We show that F is fne if and only if N is ne in the two-norm. First, we have

$$\langle Fx - Fy, x - y\rangle = \frac{1}{2}\|x - y\|_2^2 + \frac{1}{2}\langle Nx - Ny, x - y\rangle.$$

Also,

$$\|\frac{1}{2}(I+N)x - \frac{1}{2}(I+N)y\|_2^2 = \frac{1}{4}\|x-y\|_2^2 + \frac{1}{4}\|Nx - Ny\|_2^2 + \frac{1}{2}\langle Nx - Ny, x - y\rangle.$$

Therefore,

$$\langle Fx - Fy, x - y\rangle \geq \|Fx - Fy\|_2^2$$

if and only if

$$\|Nx - Ny\|_2^2 \leq \|x - y\|_2^2.$$

∎

Corollary 6.3 *For $m = 1, 2, ..., M$, let $\alpha_m > 0$, with $\sum_{m=1}^{M} \alpha_m = 1$, and let $F_m : \mathbb{R}^J \to \mathbb{R}^J$ be fne. Then the operator*

$$F = \sum_{m=1}^{M} \alpha_m F_m$$

is also fne. In particular, the arithmetic mean of the F_m is fne.

Corollary 6.4 *An operator F is fne if and only if $I - F$ is fne.*

6.6 Firmly Nonexpansive Gradients

In this section we consider some useful properties of the gradient operator of a differentiable convex function.

It is convenient for us to consider functions on $\mathbb{R}^J$ whose values may be infinite. For example, we define the *indicator function* ι_C of a set $C \subseteq \mathbb{R}^J$ to have the value zero for x in C, and the value $+\infty$ for x outside the set C.

Definition 6.12 *A function $f : \mathbb{R}^J \to [-\infty, \infty]$ is* proper *if there is no x for which $f(x) = -\infty$ and some x for which $f(x) < +\infty$.*

All the functions we shall consider in this text will be proper.

Definition 6.13 *Let f be a proper function defined on $\mathbb{R}^J$. The subset of $\mathbb{R}^{J+1}$ defined by*

$$\text{epi}(f) = \{(x, \gamma) | f(x) \leq \gamma\}$$

is the epi-graph *of f. Then we say that f is* convex *if its epi-graph is a convex set.*

Alternative definitions of convex function are presented in the exercises.

Definition 6.14 *The* effective domain *of a proper function $f : \mathbb{R}^J \to (-\infty, \infty]$ is the set*

$$\text{dom}f = \{x | f(x) < +\infty\}.$$

It is also the projection onto $\mathbb{R}^J$ of its epi-graph.

It is easily shown that the effective domain of a convex function is a convex set.

Let $g : \mathbb{R}^J \to \mathbb{R}$ be differentiable. We have several equivalent notions of convexity for such functions of several variables.

Theorem 6.2 *Let $g : \mathbb{R}^J \to \mathbb{R}$ be differentiable. The following are equivalent:*

1. *$g(x)$ is convex;*
2. *for all a and b we have*

$$g(b) \geq g(a) + \langle \nabla g(a), b - a \rangle ; \tag{6.21}$$

3. *for all a and b we have*

$$\langle \nabla g(b) - \nabla g(a), b - a \rangle \geq 0. \tag{6.22}$$

Corollary 6.5 *The function* $g(x) = \frac{1}{2}\left(\|x\|_2^2 - \|x - P_C x\|_2^2\right)$ *is convex.*

Proof: We show later in Corollary 8.1 that the gradient of $g(x)$ is $\nabla g(x) = P_C x$. From the inequality (6.15) we know that

$$\langle P_C x - P_C y, x - y\rangle \geq 0,$$

for all x and y. Therefore, $g(x)$ is convex, by Theorem 6.2. ∎

The following theorem is a consequence of the somewhat more general Baillon-Haddad Theorem (see Corollary 18.16 in [18]).

Theorem 6.3 *Let $h(x)$ be convex and differentiable and its derivative, $\nabla h(x)$, nonexpansive in the two-norm, that is,*

$$||\nabla h(b) - \nabla h(a)||_2 \leq ||b - a||_2, \tag{6.23}$$

for all a and b. Then $\nabla h(x)$ is firmly nonexpansive. which means that

$$\langle \nabla h(b) - \nabla h(a), b - a\rangle \geq ||\nabla h(b) - \nabla h(a)||_2^2. \tag{6.24}$$

Suppose that $g(x) : \mathbb{R}^J \to \mathbb{R}$ is convex and the operator ∇g is L-Lipschitz. Let $h(x) = \frac{1}{L}g(x)$, so that ∇h is a nonexpansive operator. According to Theorem 6.3, the operator $\nabla h = \frac{1}{L}\nabla g$ is firmly nonexpansive.

In [133] Golshtein and Tretyakov prove the following theorem, from which Theorem 6.3 follows immediately. The proof given here is different from that given in [133].

Theorem 6.4 *Let $g : \mathbb{R}^J \to \mathbb{R}$ be convex and differentiable. The following are equivalent:*

1.

$$||\nabla g(x) - \nabla g(y)||_2 \leq ||x - y||_2; \tag{6.25}$$

2.

$$g(x) \geq g(y) + \langle \nabla g(y), x - y\rangle + \frac{1}{2}||\nabla g(x) - \nabla g(y)||_2^2; \tag{6.26}$$

and

3.

$$\langle \nabla g(x) - \nabla g(y), x - y\rangle \geq ||\nabla g(x) - \nabla g(y)||_2^2. \tag{6.27}$$

Proof: The only non-trivial step in the proof is showing that Inequality (6.25) implies Inequality (6.26). From Theorem 6.2 we see that Inequality (6.25) implies that the function $h(x) = \frac{1}{2}\|x\|^2 - g(x)$ is convex, and that

$$\frac{1}{2}\|x-y\|^2 \geq g(x) - g(y) - \langle \nabla g(y), x-y \rangle ,$$

for all x and y. Now fix y and define

$$d(z) = D_g(z,y) = g(z) - g(y) - \langle \nabla g(y), z-y \rangle,$$

for all z. Since the function $g(z)$ is convex, so is $d(z)$. Since

$$\nabla d(z) = \nabla g(z) - \nabla g(y),$$

it follows from Inequality (6.25) that

$$\|\nabla d(z) - \nabla d(x)\| \leq \|z-x\|,$$

for all x and z. Then, from our previous calculations, we may conclude that

$$\frac{1}{2}\|z-x\|^2 \geq d(z) - d(x) - \langle \nabla d(x), z-x \rangle ,$$

for all z and x.

Now let x be arbitrary and

$$z = x - \nabla g(x) + \nabla g(y).$$

Then

$$0 \leq d(z) \leq d(x) - \frac{1}{2}\|\nabla g(x) - \nabla g(y)\|^2.$$

This completes the proof. ∎

We know from Corollary 6.5 that the function

$$g(x) = \frac{1}{2}\Big(\|x\|_2^2 - \|x - P_C x\|_2^2\Big)$$

is convex. As Corollary 8.1 tells us, its gradient is $\nabla g(x) = P_C x$. We showed in Corollary 6.2 that the operator P_C is nonexpansive by showing that it is actually firmly nonexpansive. Therefore, Theorem 6.3 can be viewed as a generalization of Corollary 6.2.

If $g(x)$ is convex and ∇g is L-Lipschitz, then $\frac{1}{L}\nabla g$ is nonexpansive, so, by Theorem 6.3, it is firmly nonexpansive. It follows that, for $\gamma > 0$, the operator

$$T = I - \gamma \nabla g \tag{6.28}$$

is averaged, whenever $0 < \gamma < \frac{2}{L}$. By the Krasnosel'skii-Mann-Opial Theorem 7.1, the iterative sequence $x^{k+1} = Tx^k = x^k - \gamma \nabla g(x^k)$ converges to a minimizer of $g(x)$, whenever minimizers exist.

6.6.1 The Search for Other Properties of P_C

The class of nonexpansive operators is too large for our purposes; the operator $T = -I$ is nonexpansive, but the sequence $\{T^k x^0\}$ does not converge, in general, even though a fixed point, $x = 0$, exists. The class of firmly nonexpansive operators is too small for our purposes. Although the convergence of the iterative sequence $\{T^k x^0\}$ to a fixed point does hold for firmly nonexpansive T, whenever fixed points exist, the product of two or more fne operators need not be fne; that is, the class of fne operators is not *closed to finite products.* This poses a problem, since, as we shall see, products of orthogonal projection operators arise in several of the algorithms we wish to consider. We need a class of operators smaller than the ne ones, but larger than the fne ones, closed to finite products, and for which the sequence of iterates $\{T^k x^0\}$ will converge, for any x^0, whenever fixed points exist. The class we shall consider is the class of *averaged* operators. In all discussion of averaged operators the norm will be the two-norm.

6.7 Exercises

Ex. 6.1 *Let $C \subseteq \mathbb{R}^J$, and let x^n, $n = 1, ..., N$ be members of C. For $n = 1, ..., N$, let $\alpha_n > 0$, with $\alpha_1 + ... + \alpha_N = 1$. Show that, if C is convex, then the convex combination*

$$\alpha_1 x^1 + \alpha_2 x^2 + ... + \alpha_N x^N$$

is in C.

Ex. 6.2 *Prove Proposition 6.1. Hint: show that the set C is convex.*

Ex. 6.3 *Show that the subset of $\mathbb{R}^J$ consisting of all vectors x with $||x||_2 =$ 1 is not convex.*

Ex. 6.4 *Let $\|x\|_2 = \|y\|_2 = 1$ and $z = \frac{1}{2}(x + y)$ in $\mathbb{R}^J$. Show that $\|z\|_2 < 1$ unless $x = y$. Show that this conclusion does not hold if the* two-norm $\|\cdot\|_2$ *is replaced by the* one-norm.

Ex. 6.5 *Show that x is an extreme point of a convex set C if and only if the subset of C obtained by removing x remains convex.*

Ex. 6.6 *Let C be the set of all vectors x in $\mathbb{R}^J$ with $\|x\|_2 \leq 1$. Let K be a subset of C obtained by removing from C any number of its members for which $\|x\|_2 = 1$. Show that K is convex. Consequently, every x in C with $\|x\|_2 = 1$ is an extreme point of C.*

Ex. 6.7 *Let A and B be nonempty, closed convex subsets of $\mathbb{R}^J$. Define the set $B - A$ to be all x in $\mathbb{R}^J$ such that $x = b - a$ for some $a \in A$ and $b \in B$. Show that $B - A$ is closed if one of the two sets is bounded. Find an example of two disjoint unbounded closed convex sets in $\mathbb{R}^2$ that get arbitrarily close to each other. Show that, for this example, $B - A$ is not closed.*

Ex. 6.8 *Let C be a convex set and $f : C \subseteq \mathbb{R}^J \to (-\infty, \infty]$. Prove that $f(x)$ is a convex function, according to Definition 6.13, if and only if, for all x and y in C, and for all $0 < \alpha < 1$, we have*

$$f(\alpha x + (1 - \alpha)y) \leq \alpha f(x) + (1 - \alpha) f(y).$$

Ex. 6.9 *Let $f : \mathbb{R}^J \to [-\infty, \infty]$. Prove that $f(x)$ is a convex function if and only if, for all $0 < \alpha < 1$, we have*

$$f(\alpha x + (1 - \alpha)y) < \alpha b + (1 - \alpha)c,$$

whenever $f(x) < b$ and $f(y) < c$.

Ex. 6.10 *Given a point s in a convex set C, where are the points x for which $s = P_C x$?*

Ex. 6.11 *Show that it is possible to have a vector $z \in \mathbb{R}^J$ such that $\langle z - x, c - z \rangle \geq 0$ for all $c \in C$, but z is not $P_C x$.*

Ex. 6.12 (Rådström Cancellation [27])

(a) Show that, for any subset S of $\mathbb{R}^N$, we have $2S \subseteq S + S$, and $2S = S + S$ if S is convex.

(b) Find three finite subsets of $\mathbb{R}$, say A, B, and C, with A not contained in B, but with the property that $A + C \subseteq B + C$. Hint: Try to find an example where the set C is $C = \{-1, 0, 1\}$.

(c) Show that, if A and B are convex in $\mathbb{R}^N$, B is closed, and C is bounded in $\mathbb{R}^N$, then $A + C \subseteq B + C$ implies that $A \subseteq B$. Hint: Note that, under these assumptions, $2A + C = A + (A + C) \subseteq 2B + C$.

Ex. 6.13 [11] *Let A and B be nonempty closed convex subsets of $\mathbb{R}^N$. For each $a \in A$ define*

$$d(a, B) = \inf_{b \in B} \|a - b\|_2,$$

and then define

$$d(A, B) = \inf_{a \in A} d(a, B).$$

Let

$$E = \{a \in A | d(a, B) = d(A, B)\},$$

and

$$F = \{b \in B | d(b, A) = d(B, A)\};$$

assume that both E and F are not empty. The displacement vector *is $v = P_K(0)$, where K is the closure of the set $B - A$. For any transformation $T : \mathbb{R}^N \to \mathbb{R}^N$, denote by $\mathrm{Fix}(T)$ the set of all $x \in \mathbb{R}^N$ such that $Tx = x$. Prove the following:*

(a) $\|v\|_2 = d(A, B)$;

(b) $E + v = F$;

(c) $E = \mathrm{Fix}(P_A P_B) = A \cap (B - v)$;

(d) $F = \mathrm{Fix}(P_B P_A) = B \cap (A + v)$;

(e) $P_B e = P_F e = e + v$, for all $e \in E$;

(f) $P_A f = P_E f = f - v$, for all $f \in F$.

Ex. 6.14 *Prove Corollary 6.1.*

Ex. 6.15 *Prove Corollary 6.2.*

Ex. 6.16 *Prove Corollary 6.4.*

Chapter 7

Averaged and Paracontractive Operators

In this chapter we discuss two classes of operators important for iterative algorithms, the averaged operators and the paracontractive operators.

7.1 Averaged Operators

The term 'averaged operator' appears in the work of Baillon, Bruck and Reich [35, 9]. There are several ways to define averaged operators. One way is suggested by Lemma 6.2.

Definition 7.1 *An operator $T : \mathbb{R}^J \to \mathbb{R}^J$ is* averaged *(av) if there is an operator N that is ne in the two-norm and $\alpha \in (0, 1)$ such that $T = (1 - \alpha)I + \alpha N$. Then we say that T is* α-averaged.

It follows from Lemma 6.2 that T is fne if and only if T is α-averaged for $\alpha = \frac{1}{2}$. Every averaged operator is ne, with respect to the two-norm, and every fne operator is av.

We can also describe averaged operators T in terms of the complement operator, $G = I - T$.

Definition 7.2 *An operator G on $\mathbb{R}^J$ is called ν-*inverse strongly monotone *(ν-ism)[133] (also called co-coercive in [99]) if there is $\nu > 0$ such that*

$$\langle Gx - Gy, x - y\rangle \geq \nu||Gx - Gy||_2^2. \tag{7.1}$$

Lemma 7.1 *An operator T is ne, with respect to the two-norm, if and only if its complement $G = I - T$ is $\frac{1}{2}$-ism, and T is fne if and only if G is 1-ism, and if and only if G is fne. Also, T is ne if and only if $F = (I+T)/2$ is fne. If G is ν-ism and $\gamma > 0$ then the operator γG is $\frac{\nu}{\gamma}$-ism.*

Lemma 7.2 *An operator T is averaged if and only if $G = I - T$ is ν-ism for some $\nu > \frac{1}{2}$. If G is $\frac{1}{2\alpha}$-ism, for some $\alpha \in (0, 1)$, then T is α-av.*

Proof: We assume first that there is $\alpha \in (0, 1)$ and ne operator N such that $T = (1 - \alpha)I + \alpha N$, and so $G = I - T = \alpha(I - N)$. Since N is ne, $I - N$ is $\frac{1}{2}$-ism and $G = \alpha(I - N)$ is $\frac{1}{2\alpha}$-ism. Conversely, assume that G is ν-ism for some $\nu > \frac{1}{2}$. Let $\alpha = \frac{1}{2\nu}$ and write $T = (1 - \alpha)I + \alpha N$ for $N = I - \frac{1}{\alpha}G$. Since $I - N = \frac{1}{\alpha}G$, $I - N$ is $\alpha\nu$-ism. Consequently $I - N$ is $\frac{1}{2}$-ism and N is ne. ∎

An averaged operator is easily constructed from a given operator N that is ne in the two-norm by taking a convex combination of N and the identity I. The beauty of the class of av operators is that it contains many operators, such as P_C, that are not originally defined in this way. As we shall see shortly, finite products of averaged operators are again averaged, so the product of finitely many orthogonal projections is av.

We present now the fundamental properties of averaged operators, in preparation for the proof that the class of averaged operators is closed to finite products.

Note that we can establish that a given operator is av by showing that there is an α in the interval $(0, 1)$ such that the operator

$$\frac{1}{\alpha}(A - (1 - \alpha)I) \tag{7.2}$$

is ne. Using this approach, we can easily show that if T is sc, then T is av.

Lemma 7.3 *Let $T = (1 - \alpha)A + \alpha N$ for some $\alpha \in (0, 1)$. If A is averaged and N is nonexpansive then T is averaged.*

Proof: Let $A = (1 - \beta)I + \beta M$ for some $\beta \in (0, 1)$ and ne operator M. Let $1 - \gamma = (1 - \alpha)(1 - \beta)$. Then we have

$$T = (1 - \gamma)I + \gamma[(1 - \alpha)\beta\gamma^{-1}M + \alpha\gamma^{-1}N]. \tag{7.3}$$

Since the operator $K = (1 - \alpha)\beta\gamma^{-1}M + \alpha\gamma^{-1}N$ is easily shown to be ne and the convex combination of two ne operators is again ne, T is averaged. ∎

Corollary 7.1 *If A and B are av and α is in the interval $[0,1]$, then the operator $T=(1-\alpha)A+\alpha B$ formed by taking the convex combination of A and B is av.*

Corollary 7.2 *Let $T=(1-\alpha)F+\alpha N$ for some $\alpha\in(0,1)$. If F is fne and N is ne then T is averaged.*

The orthogonal projection operators P_H onto hyperplanes $H=H(a,\gamma)$ are sometimes used with *relaxation*, which means that P_H is replaced by the operator

$$T=(1-\omega)I+\omega P_H, \tag{7.4}$$

for some ω in the interval $(0,2)$. Clearly, if ω is in the interval $(0,1)$, then T is av, by definition, since P_H is ne. We want to show that, even for ω in the interval $[1,2)$, T is av. To do this, we consider the operator $R_H=2P_H-I$, which is reflection through H; that is,

$$P_Hx=\frac{1}{2}(x+R_Hx), \tag{7.5}$$

for each x.

Lemma 7.4 *The operator $R_H=2P_H-I$ is an isometry; that is,*

$$||R_Hx-R_Hy||_2=||x-y||_2, \tag{7.6}$$

for all x and y, so that R_H is ne.

Lemma 7.5 *For $\omega=1+\gamma$ in the interval $[1,2)$, we have*

$$(1-\omega)I+\omega P_H=\alpha I+(1-\alpha)R_H, \tag{7.7}$$

for $\alpha=\frac{1-\gamma}{2}$; therefore, $T=(1-\omega)I+\omega P_H$ is av.

This lemma is also true when the hyperplane H is replaced by a nonempty closed convex set.

The product of finitely many ne operators is again ne, while the product of finitely many fne operators, even orthogonal projections, need not be fne. It is a helpful fact that the product of finitely many av operators is again av.

If $A=(1-\alpha)I+\alpha N$ is averaged and B is averaged then $T=AB$ has the form $T=(1-\alpha)B+\alpha NB$. Since B is av and NB is ne, it follows from Lemma 7.3 that T is averaged. Summarizing, we have

Proposition 7.1 *If A and B are averaged, then $T=AB$ is averaged.*

7.2 Gradient Operators

Another type of operator that is averaged can be derived from gradient operators. Let $g(x) : \mathbb{R}^J \to \mathbb{R}$ be a differentiable convex function and $f(x) = \nabla g(x)$ its gradient. If ∇g is nonexpansive, then, according to Theorem 6.3, ∇g is fne. If, for some $L > 0$, ∇g is L-Lipschitz, for the two-norm, that is,

$$||\nabla g(x) - \nabla g(y)||_2 \leq L||x - y||_2, \tag{7.8}$$

for all x and y, then $\frac{1}{L}\nabla g$ is ne, therefore fne, and the operator $T = I - \gamma\nabla g$ is av, for $0 < \gamma < \frac{2}{L}$. From Corollary 8.1 we know that the operators P_C are actually gradient operators; $P_C x = \nabla g(x)$ for

$$g(x) = \frac{1}{2}(\|x\|_2^2 - \|x - P_C x\|_2^2).$$

7.3 Two Useful Identities

The identities in the next two lemmas relate an arbitrary operator T to its complement, $G = I - T$, where I denotes the identity operator. These identities will allow us to transform properties of T into properties of G that may be easier to work with. A simple calculation is all that is needed to establish the following lemma.

Lemma 7.6 *Let T be an arbitrary operator T on $\mathbb{R}^J$ and $G = I - T$. Then*

$$||x - y||_2^2 - ||Tx - Ty||_2^2 = 2(\langle Gx - Gy, x - y\rangle) - ||Gx - Gy||_2^2. \tag{7.9}$$

Lemma 7.7 *Let T be an arbitrary operator T on $\mathbb{R}^J$ and $G = I - T$. Then*

$$\langle Tx - Ty, x - y\rangle - ||Tx - Ty||_2^2 =$$

$$\langle Gx - Gy, x - y\rangle - ||Gx - Gy||_2^2. \tag{7.10}$$

Proof: Use the previous lemma. ∎

7.4 The Krasnosel'skii-Mann-Opial Theorem

For any operator T that is averaged, convergence of the sequence $\{T^k x^0\}$ to a fixed point of T, whenever fixed points of T exist, is guaranteed by the Krasnosel'skii-Mann-Opial (KMO) Theorem [152, 164, 182]:

Theorem 7.1 *Let T be α-averaged, for some $\alpha \in (0,1)$. Then, for any x^0, the sequence $\{T^k x^0\}$ converges to a fixed point of T, whenever Fix(T) is nonempty.*

Proof: Let z be a fixed point of T. The identity in Equation (7.9) is the key to proving Theorem 7.1.

Using $Tz = z$ and $(I - T)z = 0$ and setting $G = I - T$ we have

$$||z - x^k||_2^2 - ||Tz - x^{k+1}||_2^2 = 2\langle Gz - Gx^k, z - x^k\rangle - ||Gz - Gx^k||_2^2. \tag{7.11}$$

Since, by Lemma 7.2, G is $\frac{1}{2\alpha}$-ism, we have

$$||z - x^k||_2^2 - ||z - x^{k+1}||_2^2 \geq (\frac{1}{\alpha} - 1)||x^k - x^{k+1}||_2^2. \tag{7.12}$$

Consequently the sequence $\{x^k\}$ is bounded, the sequence $\{||z - x^k||_2\}$ is decreasing and the sequence $\{||x^k - x^{k+1}||_2\}$ converges to zero. Let x^* be a cluster point of $\{x^k\}$. Then we have $Tx^* = x^*$, so we may use x^* in place of the arbitrary fixed point z. It follows then that the sequence $\{||x^* - x^k||_2\}$ is decreasing; since a subsequence converges to zero, the entire sequence converges to zero. The proof is complete. ∎

A version of the KMO Theorem 7.1, with variable coefficients, appears in Reich's paper [188].

An operator T is said to be *asymptotically regular* if, for any x, the sequence $\{\|T^k x - T^{k+1}x\|\}$ converges to zero. The proof of the KMO Theorem 7.1 involves showing that any averaged operator is asymptotically regular. In [182] Opial generalizes the KMO Theorem, proving that, if T is nonexpansive and asymptotically regular, then the sequence $\{T^k x\}$ converges to a fixed point of T, whenever fixed points exist, for any x.

Note that, in the KMO Theorem, we assumed that T is α-averaged, so that $G = I - T$ is ν-ism, for some $\nu > \frac{1}{2}$. But we actually used a somewhat weaker condition on G; we required only that

$$\langle Gz - Gx, z - x\rangle \geq \nu\|Gz - Gx\|^2$$

for z such that $Gz = 0$. This weaker property is called *weakly* ν-ism.

7.5 Affine Linear Operators

It may not always be easy to decide if a given operator is averaged. The class of affine linear operators provides an interesting illustration of the problem.

Definition 7.3 *An operator* $T : \mathbb{R}^J \to \mathbb{R}^I$ *is* affine linear *or just* affine *if there is an* I *by* J *matrix* B *and a fixed vector* d *so that* $Tx = Bx + d$ *for all* x.

The affine operator $Tx = Bx + d$ will be ne, sc, fne, or av precisely when the linear operator given by multiplication by the matrix B is the same.

7.5.1 The Hermitian Case

When B is Hermitian, we can determine if B belongs to these classes by examining its eigenvalues λ. We have the following theorem.

Theorem 7.2 *Let* B *be Hermitian. Then*

- *B is nonexpansive if and only if* $-1 \leq \lambda \leq 1$, *for all* λ;
- *B is averaged if and only if* $-1 < \lambda \leq 1$, *for all* λ;
- *B is a strict contraction if and only if* $-1 < \lambda < 1$, *for all* λ;
- *B is firmly nonexpansive if and only if* $0 \leq \lambda \leq 1$, *for all* λ.

Proof: The proof is left as an exercise for the reader. ∎

Affine linear operators T that arise, for instance, in splitting methods for solving systems of linear equations, generally have non-Hermitian linear part B. Deciding if such operators belong to these classes is more difficult. Instead, we can ask if the operator is *paracontractive*, with respect to some norm.

7.6 Paracontractive Operators

By examining the properties of the orthogonal projection operators P_C, we were led to the useful class of averaged operators. The orthogonal projections also belong to another useful class, the paracontractions.

Definition 7.4 *A continuous operator T is called* paracontractive *(pc), with respect to a given norm, if, for every fixed point y of T, we have*

$$||Tx - y|| < ||x - y||, \tag{7.13}$$

unless $Tx = x$.

Paracontractive operators are studied by Censor and Reich in [91].

Proposition 7.2 *The operators $T = P_C$ are paracontractive, with respect to the Euclidean norm.*

Proof: It follows from Cauchy's Inequality that

$$||P_C x - P_C y||_2 \leq ||x - y||_2,$$

with equality if and only if

$$P_C x - P_C y = \alpha(x - y),$$

for some scalar α with $|\alpha| = 1$. But, because

$$0 \leq \langle P_C x - P_C y, x - y\rangle = \alpha||x - y||_2^2,$$

it follows that $\alpha = 1$, and so

$$P_C x - x = P_C y - y.$$

∎

When we ask if a given operator T is pc, we must specify the norm. We often construct the norm specifically for the operator involved, as in Equation (7.20). To illustrate, we consider the case of affine operators whose associated matrix is diagonalizable.

7.6.1 Linear and Affine Paracontractions

Definition 7.5 *A J by J matrix B is* diagonalizable *if $\mathbb{C}^J$ has a basis of eigenvectors of B.*

Let the matrix B be diagonalizable and let the columns of V be an eigenvector basis. Then we have $V^{-1}BV = D$, where D is the diagonal matrix having the eigenvalues of B along its diagonal.

Lemma 7.8 *A square matrix B is diagonalizable if all its eigenvalues are distinct.*

Proof: Let B be J by J. Let λ_j be the eigenvalues of B, $Bx^j = \lambda_j x^j$, and $x^j \neq 0$, for $j = 1, ..., J$. Let x^m be the first eigenvector that is in the span of $\{x_j | j = 1, ..., m-1\}$. Then

$$x^m = a_1 x^1 + ... a_{m-1} x^{m-1}, \tag{7.14}$$

for some constants a_j that are not all zero. Multiply both sides by λ_m to get

$$\lambda_m x^m = a_1 \lambda_m x^1 + ... a_{m-1} \lambda_m x^{m-1}. \tag{7.15}$$

From

$$\lambda_m x^m = A x^m = a_1 \lambda_1 x^1 + ... a_{m-1} \lambda_{m-1} x^{m-1}, \tag{7.16}$$

it follows that

$$a_1(\lambda_m - \lambda_1) x^1 + ... + a_{m-1}(\lambda_m - \lambda_{m-1}) x^{m-1} = 0, \tag{7.17}$$

from which we can conclude that some x^n in $\{x^1, ..., x^{m-1}\}$ is in the span of the others. This is a contradiction. ∎

We see from this Lemma that almost all square matrices B are diagonalizable, in the sense that, if the entries of B are chosen randomly in $\mathbb{C}$ and independently of one another, then the probability that the matrix B will have distinct eigenvalues is one. Indeed, all Hermitian B are diagonalizable. If B has real entries, but is not symmetric, then the eigenvalues of B need not be real, and the eigenvectors of B can have non-real entries. Consequently, we must consider B as a linear operator on $\mathbb{C}^J$, if we are to talk about diagonalizability. For example, consider the real matrix

$$B = \begin{bmatrix} 0 & 1 \\ -1 & 0 \end{bmatrix}. \tag{7.18}$$

Its eigenvalues are $\lambda = i$ and $\lambda = -i$. The corresponding eigenvectors are $(1, i)^T$ and $(1, -i)^T$. The matrix B is then diagonalizable as an operator on C^2, but not as an operator on $\mathbb{R}^2$.

When B is not Hermitian, it is not as easy to determine if the affine operator T is sc with respect to a given norm. Instead, we often tailor the norm to the operator T. Suppose that B is a diagonalizable matrix, that is, there is a basis for $\mathbb{C}^J$ consisting of eigenvectors of B. Let $\mathcal{A} = \{u^1, ..., u^J\}$ be such a basis, and let $Bu^j = \lambda_j u^j$, for each $j = 1, ..., J$. For each x in $\mathbb{C}^J$, there are unique coefficients a_j so that

$$x = \sum_{j=1}^{J} a_j u^j. \tag{7.19}$$

Then let

$$||x|| = \sum_{j=1}^{J} |a_j|. \tag{7.20}$$

The expression $||\cdot||$ in Equation (7.20) defines a norm on $\mathbb{C}^J$. If $\rho(B) < 1$, then the affine operator T is sc, with respect to this norm.

Proposition 7.3 *Let T be an affine linear operator whose linear part B is diagonalizable, and $|\lambda| < 1$ for all eigenvalues λ of B that are not equal to one. Then the operator T is pc, with respect to the norm given by Equation (7.20).*

Proof: This is Exercise 7.3. ∎

We see from Proposition 7.3 that, for the case of affine operators T whose linear part is not Hermitian, instead of asking if T is av, we can ask if T is pc; since B will almost certainly be diagonalizable, we can answer this question by examining the eigenvalues of B.

Unlike the class of averaged operators, the class of paracontractive operators is not necessarily closed to finite products, unless those factor operators have a common fixed point.

7.6.2 The Elsner-Koltracht-Neumann Theorem

Our interest in paracontractions is due to the Elsner-Koltracht-Neumann (EKN) Theorem [118]:

Theorem 7.3 *Let T be pc with respect to some vector norm. If T has a fixed point, then the sequence $\{T^k x^0\}$ converges to a fixed point of T, for all starting vectors x^0.*

We follow the development in [118].

Theorem 7.4 *Suppose that there is a vector norm on $\mathbb{C}^J$, with respect to which each T_i is a pc operator, for $i = 1, ..., I$, and that $F = \cap_{i=1}^I \mathrm{Fix}(T_i)$ is not empty. For $k = 0, 1, ...,$ let $i(k) = k(\bmod I) + 1$, and $x^{k+1} = T_{i(k)} x^k$. The sequence $\{x^k\}$ converges to a member of F, for every starting vector x^0.*

Proof: Let $y \in F$. Then, for $k = 0, 1, ...,$

$$||x^{k+1} - y|| = ||T_{i(k)} x^k - y|| \leq ||x^k - y||, \tag{7.21}$$

so that the sequence $\{||x^k - y||\}$ is decreasing; let $d \geq 0$ be its limit. Since the sequence $\{x^k\}$ is bounded, we select an arbitrary cluster point, x^*. Then $d = ||x^* - y||$, from which we can conclude that

$$||T_i x^* - y|| = ||x^* - y||, \tag{7.22}$$

and $T_i x^* = x^*$, for $i = 1, ..., I$; therefore, $x^* \in F$. Replacing y, an arbitrary member of F, with x^*, we have that $||x^k - x^*||$ is decreasing. But, a subsequence converges to zero, so the whole sequence must converge to zero. This completes the proof. ∎

Corollary 7.3 *If there is a vector norm on $\mathbb{C}^J$, with respect to which each T_i is a pc operator, for $i = 1, ..., I$, $T = T_I T_{I-1} \cdots T_2 T_1$, and $F = \cap_{i=1}^I \text{Fix}(T_i)$ is not empty, then $F = \text{Fix}(T)$.*

Proof: The sequence $x^{k+1} = T_{i(k)} x^k$ converges to a member of Fix(T), for every x^0. Select x^0 in F. ∎

Corollary 7.4 *The product T of two or more pc operators T_i, $i = 1, ..., I$ is again a pc operator, if $F = \cap_{i=1}^I \text{Fix}(T_i)$ is not empty.*

Proof: Suppose that for $T = T_I T_{I-1} \cdots T_2 T_1$, and $y \in F = \text{Fix}(T)$, we have

$$||Tx - y|| = ||x - y||. \tag{7.23}$$

Then, since

$$||T_I(T_{I-1} \cdots T_1)x - y|| \leq ||T_{I-1} \cdots T_1 x - y|| \leq ...$$

$$\leq ||T_1 x - y|| \leq ||x - y||, \tag{7.24}$$

it follows that

$$||T_i x - y|| = ||x - y||, \tag{7.25}$$

and $T_i x = x$, for each i. Therefore, $Tx = x$. ∎

7.7 Exercises

Ex. 7.1 *Show that, if the operator T is α-av and $1 > \beta > \alpha$, then T is β-av.*

Ex. 7.2 *Prove Lemma 7.1.*

Ex. 7.3 *Prove Proposition 7.3.*

Ex. 7.4 *Show that, if B is a linear av operator, then $|\lambda| < 1$ for all eigenvalues λ of B that are not equal to one.*

Ex. 7.5 *We say that an operator* $R : \mathbb{C}^J \to \mathbb{C}^J$ *is* quasi-averaged *if, for some operator* Q *that is qne, with respect to the two-norm, and for some* α *in the interval* $(0, 1)$, *we have*

$$R = (1 - \alpha)I + \alpha Q.$$

Show that the Krasnosel'skii-Mann-Opial Theorem 7.1 holds when averaged operators are replaced by quasi-averaged operators.

Chapter 8

Convex Feasibility and Related Problems

8.1 Convex Constraint Sets

When we minimize a real-valued function $f(x)$, constraints on x often take the form of inclusion in certain convex sets. These sets may be related to the measured data, or incorporate other aspects of x known *a priori*. There are several related problems that then arise. Iterative algorithms based on orthogonal projection onto convex sets are then employed to solve these problems. Such constraints can often be formulated as requiring that the desired x lie within the intersection C of a finite collection $\{C_1, ..., C_I\}$ of convex sets.

8.1.1 Convex Feasibility

When the number of convex sets is large and the intersection C small, any member of C may be sufficient for our purposes. Finding such x is the *convex feasibility problem* (CFP).

8.1.2 Constrained Optimization

When the intersection C is large, simply obtaining an arbitrary member of C may not be enough; we may require, in addition, that the chosen x optimize some cost function. For example, we may seek the x in C that minimizes $||x - x^0||_2^2$. This is *constrained optimization.*

8.1.3 Proximity Function Minimization

When the collection of convex sets has empty intersection, we may minimize a *proximity function*, such as

$$f(x) = \frac{1}{2I} \sum_{i=1}^{I} ||P_{C_i}x - x||_2^2. \tag{8.1}$$

When the set C is non-empty, the smallest value of $f(x)$ is zero, and is attained at any member of C. When C is empty, the minimizers of $f(x)$, when they exist, provide a reasonable approximate solution to the CFP.

8.1.4 The Split-Feasibility Problem

An interesting variant of the CFP is the *split-feasibility problem* (SFP) [83]. Let A be an I by J real matrix. The SFP is to find a member of a closed, convex set C in $\mathbb{R}^J$ for which Ax is a member of a second closed, convex set Q in $\mathbb{R}^I$. When there is no such x, we can obtain an approximate solution by minimizing the proximity function

$$f(x) = \frac{1}{2}||P_Q Ax - Ax||_2^2, \tag{8.2}$$

over all x in C, whenever such minimizers exist.

8.1.5 Differentiability

The following theorem describes the gradient of the function $f(x)$ in Equation (8.2).

Theorem 8.1 *Let $f(x) = \frac{1}{2}||P_Q Ax - Ax||_2^2$ and $t \in \partial f(x)$. Then $t = A^T(I - P_Q)Ax$, so that $t = \nabla f(x)$.*

Proof: First, we show that $t = A^T z^*$ for some z^*. Let $s = x + w$, where w is an arbitrary member of the null space of A. Then $As = Ax$ and $f(s) = f(x)$. From

$$0 = f(s) - f(x) \geq \langle t, s - x \rangle = \langle t, w \rangle,$$

it follows that

$$\langle t, w \rangle = 0,$$

for all w in the null space of A, from which we conclude that t is in the range of A^T. Therefore, we can write $t = A^T z^*$.

Let u be chosen so that $\|A(u - x)\| = 1$, and let $\epsilon > 0$. We then have

$$\|P_Q Ax - A(x + \epsilon(u - x))\|^2 - \|P_Q Ax - Ax\|^2 \geq$$

$$\|P_Q(Ax + \epsilon(u - x)) - A(x + \epsilon(u - x))\|^2 - \|P_Q Ax - Ax\|^2 \geq 2\epsilon\langle t, u - x\rangle.$$

Therefore, since

$$\|P_Q Ax - A(x + \epsilon(u - x))\|^2 = \|P_Q Ax - Ax\|^2 - 2\epsilon\langle P_Q Ax - Ax, A(u - x)\rangle + \epsilon^2,$$

it follows that

$$\frac{\epsilon}{2} \geq \langle P_Q Ax - Ax + z^*, A(u - x)\rangle = -\langle A^T(I - P_Q)Ax - t, u - x\rangle.$$

Since ϵ is arbitrary, it follows that

$$\langle A^T(I - P_Q)Ax - t, u - x\rangle \geq 0,$$

for all appropriate u. But this is also true if we replace u with $v = 2x - u$. Consequently, we have

$$\langle A^T(I - P_Q)Ax - t, u - x\rangle = 0.$$

Now we select

$$u - x = (A^T(I - P_Q)Ax - t)/\|AA^T(I - P_Q)Ax - At\|,$$

from which it follows that

$$A^T(I - P_Q)Ax = t.$$

∎

Corollary 8.1 *The gradient of the function*

$$f(x) = \frac{1}{2}\|x - P_C x\|^2$$

is $\nabla f(x) = x - P_C x$, and the gradient of the function

$$g(x) = \frac{1}{2}\left(\|x\|_2^2 - \|x - P_C x\|_2^2\right)$$

is $\nabla g(x) = P_C x$.

Just as the function $h(t) = t^2$ is differentiable for all real t, but the function $f(t) = |t|$ is not differentiable at $t = 0$, the function

$$h_0(x) = \|x\|_2$$

is not differentiable at $x = 0$ and the function

$$h(x) = \|x - P_C x\|_2$$

is not differentiable at boundary points of the set C.

If C is a nonempty closed convex subset of $\mathbb{R}^J$, then $N_C(x)$, the *normal cone* to C at x, is the empty set, if x is not a member of C, and if $x \in C$, then

$$N_C(x) = \{u | \langle u, c - x\rangle \leq 0, \text{for all}\, c \in C\}. \tag{8.3}$$

We have the following theorem.

Theorem 8.2 *For any x in the interior of C, the gradient of the function $h(x) = \|x - P_C x\|_2$ is $\nabla h(x) = 0$. For any x outside C the gradient is*

$$\nabla h(x) = \frac{x - P_C x}{\|x - P_C x\|_2}.$$

For x on the boundary of C, however, the function $h(x)$ is not differentiable; any vector u in the normal cone $N_C(x)$ with $\|u\|_2 \leq 1$ is in $\partial h(x)$.

Proof: The function $g(t) = \sqrt{t}$ is differentiable for all positive values of t, so the function

$$h(x) = \left(\|x - P_C x\|_2^2\right)^{1/2}$$

is differentiable whenever x is not in C. Using the Chain Rule, we get

$$\nabla h(x) = \frac{x - P_C x}{\|x - P_C x\|_2}.$$

For x in the interior of C, the function $h(x)$ is identically zero in a neighborhood of x, so that the gradient is zero there. The only difficult case is when x is on the boundary of C.

First, we assume that $u \in N_C(x)$ and $\|u\|_2 = 1$. Then we must show that

$$\langle u, y - x\rangle \leq \|y - P_C y\|_2.$$

If y is such that the inner product is non-positive, then the inequality is clearly true. So we focus on those y for which the inner product is positive, which means that y lies in the half-space bounded by the hyperplane H, where

$$H = \{z | \langle u, z\rangle \geq \langle u, x\rangle\}.$$

The vector $y - P_H y$ is the orthogonal projection of the vector $y - x$ onto the line containing y and $P_H y$, which also contains the vector u. Therefore,

$$y - P_H y = \langle u, y - x\rangle u,$$

and

$$\|y - P_C y\|_2 \geq \|y - P_H y\|_2 = \langle u, y - x\rangle.$$

Now we prove the converse.

We assume now that

$$\langle u, y - x\rangle \leq \|y - P_C y\|_2,$$

for all y, and show that $\|u\|_2 \leq 1$ and $u \in N_C(x)$. If u is not in $N_C(x)$, then there is a $y \in C$ with

$$\langle u, y - x\rangle > 0,$$

but $\|y - P_C y\|_2 = 0$. Finally, we must show that $\|u\|_2 \leq 1$.

Let $y = x + u$, so that $P_C y = x$. Then

$$\langle u, y - x\rangle = \langle u, u\rangle = \|u\|_2^2,$$

while

$$\|y - P_C y\|_2 = \|y - x\|_2 = \|u\|_2.$$

It follows that $\|u\|_2 \leq 1$. ∎

We are used to thinking of functions that are not differentiable as lacking something. From the point of view of subgradients, not being differentiable means having too many of something.

The gradient of the function $f(x)$ in Equation (8.1) is

$$\nabla f(x) = x - \frac{1}{I}\sum_{i=1}^{I} P_{C_i}x. \tag{8.4}$$

Therefore, a gradient descent approach to minimizing $f(x)$ has the iterative step

$$x^{k+1} = x^k - \gamma_k\Big(x^k - \frac{1}{I}\sum_{i=1}^{I} P_{C_i}x^k\Big) =$$

$$(1-\gamma_k)x^k + \gamma_k\Big(\frac{1}{I}\sum_{i=1}^{I} P_{C_i}x^k\Big). \tag{8.5}$$

This is sometimes called the *relaxed averaged projections algorithm.* As we shall see shortly, the choice of $\gamma_k = 1$ is sufficient for convergence.

8.2 Using Orthogonal Projections

When the convex sets are half-spaces in two or three dimensional space, we may be able to find a member of their intersection by drawing a picture or just by thinking; in general, however, solving the CFP must be left up to the computer and we need an algorithm.

8.2.1 Successive Orthogonal Projection

The CFP can be solved using the *successive orthogonal projections* (SOP) method.

Algorithm 8.1 (SOP) *For arbitrary x^0, let*

$$x^{k+1} = P_I P_{I-1}\cdots P_2 P_1 x^k, \tag{8.6}$$

where $P_i = P_{C_i}$ is the orthogonal projection onto C_i.

For non-empty C, convergence of the SOP to a solution of the CFP will follow, once we have established that, for any x^0, the iterative sequence $\{T^k x^0\}$ converges to a fixed point of T, where

$$T = P_I P_{I-1}\cdots P_2 P_1. \tag{8.7}$$

Lemma 8.1 *The fixed points of T are the members of the set C.*

Proof: Clearly every member of C is a fixed point of T. Suppose that $Tx = x$. Let c be a member of C. Then

$$\|x - c\|_2 = \|Tx - c\|_2 \leq \|P_1 x - c\| - 2 \leq \|x - c\|_2.$$

Therefore, $P_1 x = x$, since P_1 is paracontractive in the two-norm. Similar arguments show that $P_i x = x$ for each i. ∎

Since T is an averaged operator, the convergence of the SOP to a member of C follows from the KMO Theorem 7.1, provided C is non-empty.

The SOP is useful when the sets C_i are easily described and the P_i are easily calculated, but P_C is not. The SOP converges to the member of C closest to x^0 when the C_i are hyperplanes, but not in general.

A good illustration of the SOP method is the *algebraic reconstruction technique* (ART) [134], also known as Kaczmarz's algorithm [148]. Associated with the system of linear equations $Ax = b$ are the hyperplanes $H_i \subseteq \mathbb{R}^J$ defined by

$$H_i = \{x|(Ax)_i = b_i\},$$

for $i = 1, ..., I$. At the kth step of the ART we get x^{k+1} by projecting the current vector x^k orthogonally onto H_i, for $i = k \,(\text{mod}\, I)$. We discuss the ART in more detail in a subsequent section.

8.2.2 Simultaneous Orthogonal Projection

When $C = \cap_{i=1}^I C_i$ is empty and we seek to minimize the proximity function $f(x)$ in Equation (8.1), we can use the simultaneous orthogonal projections (SIMOP) approach:

Algorithm 8.2 (SIMOP) *For arbitrary x^0, let*

$$x^{k+1} = \frac{1}{I}\sum_{i=1}^{I} P_i x^k. \tag{8.8}$$

The operator

$$T = \frac{1}{I}\sum_{i=1}^{I} P_i \tag{8.9}$$

is also averaged, so this iteration converges, by Theorem 7.1, whenever $f(x)$ has a minimizer.

When the convex sets are the hyperplanes $C_i = H_i$ the iteration in Equation (8.8) becomes Cimmino's algorithm, which can be written as

$$x^{k+1} = x^k - \frac{1}{I}A^T(Ax^k - b), \tag{8.10}$$

if the rows of A are first rescaled to have Euclidean length one. The more general Landweber algorithm has the iterative step

$$x^{k+1} = x^k - \gamma A^T(Ax^k - b), \tag{8.11}$$

for γ in the interval $(0, 2/\rho(A^TA))$.

8.2.3 Estimating the Spectral Radius

As we just saw, the step-length parameter γ in the Landweber algorithm is bounded above by a quantity that involves the spectral radius of the matrix A^TA. This is also true of the CQ algorithm. This poses certain difficulties. We usually resort to using iterative methods because the matrix A is too large to use anything else. In such cases, even calculating the matrix A^TA is out of the question. We need to obtain decent estimates of $\rho(A^TA)$ that are based solely on A itself. In many remote-sensing applications, such as transmission and emission tomography, the matrix A is sparse, meaning that most of its entries are zero. When we form A^TA we lose the sparseness. We would like estimates of $\rho(A^TA)$ that employ only A and are particularly useful when A is sparse.

The 1-norm of a matrix A is

$$\|A\|_1 = \max\{\sum_{i=1}^{I} |A_{ij}|, j = 1, 2, ..., J\}. \tag{8.12}$$

The infinity norm of a matrix A is

$$\|A\|_\infty = \max\{\sum_{j=1}^{J} |A_{ij}|, i = 1, 2, ..., I\}. \tag{8.13}$$

It is easily shown that

$$\rho(A^TA) \leq \|A\|_1 \|A\|_\infty,$$

so that, when $|A_{ij}| \leq 1$ for all i and j, we can say

$$\rho(A^TA) \leq IJ.$$

But we can do better than this. Suppose that $\|A\|_\infty \leq 1$. Then

$$\rho(A^TA) \leq I.$$

Suppose, in addition, that A is sparse and s is the maximum number of non-zero entries in any column of A. Then

$$\rho(A^TA) \leq s.$$

In Chapter 9 we improve this upper bound on $\rho(A^TA)$ by showing that, when the rows of A are normalized to have Euclidean length one, we again have $\rho(A^TA) \leq s$.

8.2.4 The CQ Algorithm for the SFP

The CQ algorithm is an iterative method for solving the SFP [58, 59].

Algorithm 8.3 (CQ) *For arbitrary x^0, let*

$$x^{k+1} = P_C(x^k - \gamma A^T(I - P_Q)Ax^k). \tag{8.14}$$

The operator

$$T = P_C(I - \gamma A^T(I - P_Q)A) \tag{8.15}$$

is averaged whenever γ is in the interval $(0, 2/L)$, where L is the largest eigenvalue of A^TA, and so the CQ algorithm converges to a fixed point of T, whenever such fixed points exist. When the SFP has a solution, the CQ algorithm converges to a solution; when it does not, the CQ algorithm converges to a minimizer, over C, of the proximity function $f(x) = \frac{1}{2}||P_QAx - Ax||_2^2$, whenever such minimizers exist. The function $f(x)$ is convex and, according to Theorem 8.1, its gradient is

$$\nabla f(x) = A^T(I - P_Q)Ax. \tag{8.16}$$

The convergence of the CQ algorithm then follows from Theorem 7.1. In [101] Combettes and Wajs use proximity operators to generalize the CQ algorithm.

Multi-set generalizations of the CQ algorithm have been applied recently to problems in intensity-modulated radiation therapy [81, 85].

8.2.5 An Extension of the CQ Algorithm

Let $C \in \mathbb{R}^N$ and $Q \in \mathbb{R}^M$ be closed, non-empty convex sets, and let A and B be J by N and J by M real matrices, respectively. The problem is to find $x \in C$ and $y \in Q$ such that $Ax = By$. When there are no such x and y, we consider the problem of minimizing

$$f(x, y) = \frac{1}{2}\|Ax - By\|_2^2,$$

over $x \in C$ and $y \in Q$.

Let $K = C \times Q$ in $\mathbb{R}^N \times \mathbb{R}^M$. Define

$$G = \begin{bmatrix} A & -B \end{bmatrix},$$

and

$$w = \begin{bmatrix} x \\ y \end{bmatrix},$$

so that

$$G^TG = \begin{bmatrix} A^TA & -A^TB \\ -B^TA & B^TB \end{bmatrix}.$$

The original problem can now be reformulated as finding $w \in K$ with $Gw = 0$. We shall consider the more general problem of minimizing the function $\|Gw\|$ over $w \in K$. The projected Landweber algorithm (PLW) solves this more general problem.

The iterative step of the PLW algorithm is the following:

$$w^{k+1} = P_K(w^k - \gamma G^T(Gw^k)). \tag{8.17}$$

Expressing this in terms of x and y, we obtain

$$x^{k+1} = P_C(x^k - \gamma A^T(Ax^k - By^k)); \tag{8.18}$$

and

$$y^{k+1} = P_Q(y^k + \gamma B^T(Ax^k - By^k)). \tag{8.19}$$

The PLW converges, in this case, to a minimizer of $\|Gw\|$ over $w \in K$, whenever such minimizers exist, for $0 < \gamma < \frac{2}{\rho(G^TG)}$.

8.2.6 Projecting onto the Intersection of Convex Sets

When the intersection $C = \cap_{i=1}^I C_i$ is large, and just finding any member of C is not sufficient for our purposes, we may want to calculate the orthogonal projection of x^0 onto C using the operators P_{C_i}. We cannot use the SOP unless the C_i are hyperplanes; instead we can use Dykstra's algorithm [115] or the Halpern-Lions-Wittmann-Bauschke (HLWB) algorithm; see below. There are other more general results along these lines; see [79, 144, 212, 213].

Dykstra's algorithm employs the projections P_{C_i}, but not directly on x^k, but on translations of x^k. It is motivated by the following lemma:

Lemma 8.2 *If $x = c + \sum_{i=1}^I p_i$, where, for each i, $c = P_{C_i}(c + p_i)$, then $c = P_C x$.*

Proof: The proof is left as Exercise 8.1. ∎

8.2.6.1 Dykstra's Algorithm

Dykstra's algorithm, for the simplest case of two convex sets A and B, is the following:

Algorithm 8.4 (Dykstra) *Let $b_0 = x$, and $p_0 = q_0 = 0$. Then let*

$$a_n = P_A(b_{n-1} + p_{n-1}), \tag{8.20}$$

$$b_n = P_B(a_n + q_{n-1}), \tag{8.21}$$

and define p_n and q_n by

$$x = a_n + p_n + q_{n-1} = b_n + p_n + q_n. \tag{8.22}$$

Using the algorithm, we construct two sequences, $\{a_n\}$ and $\{b_n\}$, both converging to $c = P_C x$, along with two other sequences, $\{p_n\}$ and $\{q_n\}$. Usually, but not always, $\{p_n\}$ converges to p and $\{q_n\}$ converges to q, so that

$$x = c + p + q, \tag{8.23}$$

with

$$c = P_A(c + p) = P_B(c + q). \tag{8.24}$$

Generally, however, $\{p_n + q_n\}$ converges to $x - c$.

8.2.6.2 The Halpern-Lions-Wittmann-Bauschke Algorithm

There is yet another approach to finding the orthogonal projection of the vector x onto the nonempty intersection C of finitely many closed, convex sets C_i, $i = 1, ..., I$.

Algorithm 8.5 (HLWB) *Let x^0 be arbitrary. Then let*

$$x^{k+1} = t_k x + (1 - t_k) P_{C_i} x^k, \tag{8.25}$$

where P_{C_i} denotes the orthogonal projection onto C_i, t_k is in the interval $(0, 1)$, and $i = k(\text{mod}\, I) + 1$.

Several authors have proved convergence of the sequence $\{x^k\}$ to $P_C x$, with various conditions imposed on the parameters $\{t_k\}$. As a result, the algorithm is known as the Halpern-Lions-Wittmann-Bauschke (HLWB) algorithm, after the names of several who have contributed to the evolution of the theorem; see also Corollary 2 in Reich's paper [189]. The conditions imposed by Bauschke [10] are $\{t_k\} \to 0$, $\sum t_k = \infty$, and $\sum |t_k - t_{k+I}| < +\infty$. The HLWB algorithm has been extended by Deutsch and Yamada [110] to minimize certain (possibly non-quadratic) functions over the intersection of fixed point sets of operators more general than P_{C_i}. Bregman discovered an iterative algorithm for minimizing a more general convex function $f(x)$ over x with $Ax = b$ and also x with $Ax \geq b$ [31]. These algorithms are based on his extension of the SOP to include projections with respect to generalized distances, such as entropic distances.

8.3 The ART

Let A be a complex matrix with I rows and J columns, and let b be a member of $\mathbb{C}^I$. We want to solve the system $Ax = b$. For each index value

i, let H_i be the hyperplane of J-dimensional vectors given by

$$H_i = \{x|(Ax)_i = b_i\}, \tag{8.26}$$

and P_i the orthogonal projection operator onto H_i. Let x^0 be arbitrary and, for each nonnegative integer k, let $i(k) = k(\text{mod}\, I) + 1$. The iterative step of the ART is

$$x^{k+1} = P_{i(k)}x^k. \tag{8.27}$$

Because the ART uses only a single equation at each step, it has been called a *row-action* method .

8.3.1 Calculating the ART

Given any vector z, the vector in H_i closest to z, in the sense of the Euclidean distance, has the entries

$$x_j = z_j + \overline{A_{ij}}(b_i - (Az)_i)/\sum_{m=1}^{J}|A_{im}|^2. \tag{8.28}$$

To simplify our calculations, we shall assume, throughout this chapter, that the rows of A have been rescaled to have Euclidean length one; that is

$$\sum_{j=1}^{J}|A_{ij}|^2 = 1, \tag{8.29}$$

for each $i = 1, ..., I$, and that the entries of b have been rescaled accordingly, to preserve the equations $Ax = b$. The ART is then the following: begin with an arbitrary vector x^0; for each nonnegative integer k, having found x^k, the next iterate x^{k+1} has entries

$$x_j^{k+1} = x_j^k + \overline{A_{ij}}(b_i - (Ax^k)_i). \tag{8.30}$$

As we shall show, when the system $Ax = b$ has exact solutions the ART converges to the solution closest to x^0, in the 2-norm. How fast the algorithm converges will depend on the ordering of the equations and on whether or not we use relaxation. In selecting the equation ordering, the important thing is to avoid particularly bad orderings, in which the hyperplanes H_i and H_{i+1} are nearly parallel.

8.3.2 Full-cycle ART

We again consider the *full-cycle* ART, with iterative step $z^{m+1} = Tz^m$, for

$$T = P_I P_{I-1} \cdots P_2 P_1. \tag{8.31}$$

When the system $Ax = b$ has solutions, the fixed points of T are solutions. When there are no solutions of $Ax = b$, the operator T will still have fixed points, but they will no longer be exact solutions.

8.3.3 The Basic Convergence Theorem

For a positive integer N with $1 \leq N \leq I$, we let $B_1, ..., B_N$ be not necessarily disjoint subsets of the set $\{i = 1, ..., I\}$; the subsets B_n are called *blocks*. We then let A_n be the matrix and b^n the vector obtained from A and b, respectively, by removing all the rows except for those whose index i is in the set B_n. For each n, we let s_{nt} be the number of non-zero entries in the tth column of the matrix A_n, s_n the maximum of the s_{nt}, s the maximum of the s_n, and $L_n = \rho(A_n^\dagger A_n)$ be the spectral radius, or largest eigenvalue, of the matrix $A_n^\dagger A_n$, with $L = \rho(A^\dagger A)$. We denote by A_i the ith row of the matrix A, and by ν_i the length of A_i, so that

$$\nu_i^2 = \sum_{j=1}^{J} |A_{ij}|^2.$$

The following theorem is a basic convergence result concerning block-iterative ART algorithms.

Theorem 8.3 *Let $L_n \leq 1$, for $n = 1, 2, ..., N$. If the system $Ax = b$ is consistent, then, for any starting vector x^0, and with $n = n(k) = k(\text{mod}\, N)$ and $\lambda_k \in [\epsilon, 2 - \epsilon]$ for all k, the sequence $\{x^k\}$ with iterative step*

$$x^k = x^{k-1} + \lambda_k A_n^\dagger (b^n - A_n x^{k-1}) \tag{8.32}$$

converges to the solution of $Ax = b$ for which $\|x - x^0\|_2$ is minimized.

Proof: Let $Az = b$. Applying Equation (7.9) to the operator

$$Tx = x + \lambda_k A_n^\dagger (b^n - A_n x),$$

we obtain

$$\|z - x^{k-1}\|_2^2 - \|z - x^k\|_2^2 = 2\lambda_k \|b^n - A_n x^{k-1}\|_2^2 - \lambda_k^2 \|A_n^\dagger b^n - A_n^\dagger A_n x^{k-1}\|_2^2. \tag{8.33}$$

Since $L_n \leq 1$, it follows that

$$\|A_n^\dagger b^n - A_n^\dagger A_n x^{k-1}\|_2^2 \leq \|b^n - A_n x^{k-1}\|_2^2.$$

Therefore,

$$\|z - x^{k-1}\|_2^2 - \|z - x^k\|_2^2 \geq (2\lambda_k - \lambda_k^2) \|b^n - A_n x^{k-1}\|_2^2,$$

from which we draw several conclusions:

- the sequence $\{\|z - x^k\|_2\}$ is decreasing;
- the sequence $\{\|b^n - A_n x^{k-1}\|_2\}$ converges to zero.

In addition, for fixed $n = 1, ..., N$ and $m \to \infty$,

- the sequence $\{\|b^n - A_n x^{mN+n-1}\|_2\}$ converges to zero;
- the sequence $\{x^{mN+n}\}$ is bounded.

Let $x^{*,1}$ be a cluster point of the sequence $\{x^{mN+1}\}$; then there is subsequence $\{x^{m_r N+1}\}$ converging to $x^{*,1}$. The sequence $\{x^{m_r N+2}\}$ is also bounded, and we select a cluster point $x^{*,2}$. Continuing in this fashion, we obtain cluster points $x^{*,n}$, for $n = 1, ..., N$. From the conclusions reached previously, we can show that $x^{*,n} = x^{*,n+1} = x^*$, for $n = 1, 2, ..., N-1$, and $Ax^* = b$. Replacing the generic solution $\hat{x}$ with the solution x^*, we see that the sequence $\{\|x^* - x^k\|_2\}$ is decreasing. But, subsequences of this sequence converge to zero, so the entire sequence converges to zero, and so $x^k \to x^*$.

Now we show that x^* is the solution of $Ax = b$ that minimizes $\|x - x^0\|_2$. Since $x^k - x^{k-1}$ is in the range of $A^\dagger$ for all k, so is $x^* - x^0$, from which it follows that x^* is the solution minimizing $\|x - x^0\|_2$. Another way to get this result is to use Equation (8.33). Since the right side of Equation (8.33) is independent of the choice of solution, so is the left side. Summing both sides over the index k reveals that the difference

$$\|x - x^0\|_2^2 - \|x - x^*\|_2^2$$

is independent of the choice of solution. Consequently, minimizing $\|x - x^0\|_2$ over all solutions x is equivalent to minimizing $\|x - x^*\|_2$ over all solutions x; the solution to the latter problem is clearly $x = x^*$. ∎

8.3.4 Relaxed ART

The ART employs orthogonal projections onto the individual hyperplanes. If we permit the next iterate to fall short of the hyperplane, or somewhat beyond it, we get a relaxed version of ART.The relaxed ART algorithm is as follows:

Algorithm 8.6 (Relaxed ART) *With* $\omega \in (0, 2)$, x^0 *arbitrary, and* $i = k(\text{mod}\, I) + 1$, *let*

$$x_j^{k+1} = x_j^k + \omega \overline{A_{ij}}(b_i - (Ax^k)_i)). \tag{8.34}$$

The relaxed ART converges to the solution closest to x^0, in the consistent case. In the inconsistent case, it does not converge, but subsequences associated with the same i converge to distinct vectors, forming a limit cycle.

8.3.5 Constrained ART

Let C be a closed, nonempty convex subset of $\mathbb{C}^J$ and $P_C x$ the orthogonal projection of x onto C. If there are solutions of $Ax = b$ that lie within C, we can find them using the constrained ART algorithm:

Algorithm 8.7 (Constrained ART) *With x^0 arbitrary and $i = k(\mathrm{mod}\, I) + 1$, let*

$$z_j^{k+1} = x_j^k + \overline{A_{ij}}(b_i - (Ax^k)_i), \tag{8.35}$$

and $x^{k+1} = P_C z^{k+1}$.

For example, if A and b are real and we seek a nonnegative solution to $Ax = b$, we can use the nonnegative ART algorithm:

Algorithm 8.8 (nonnegative ART) *With A a real matrix, $i = k(\mathrm{mod}\, I) + 1$, and x^0 arbitrary, let*

$$x_j^{k+1} = (x_j^k + A_{ij}(b_i - (Ax^k)_i))_+, \tag{8.36}$$

where, for any real number a, $a_+ = \max\{a, 0\}$.

The constrained ART converges to a solution of $Ax = b$ within C, whenever such solutions exist.

8.3.6 When $Ax = b$ Has No Solutions

When there are no exact solutions, the ART does not converge to a single vector, but, for each fixed i, the subsequence $\{x^{nI+i}, n = 0, 1, ...\}$ converges to a vector z^i and the collection $\{z^i \,|i = 1, ..., I\}$ is called the *limit cycle*. This was shown by Tanabe [202] and also follows from the results of De Pierro and Iusem [109]. Proofs of subsequential convergence are given in [61, 62]. The ART limit cycle will vary with the ordering of the equations, and contains more than one vector unless an exact solution exists.

Open Question: If there is a unique geometric least-squares solution, where is it, in relation to the vectors of the limit cycle? Can it be calculated easily, from the vectors of the limit cycle?

There is a partial answer to the second question. In [51] (see also [61]) it was shown that if the system $Ax = b$ has no exact solution, and if $I = J+1$, then the vectors of the limit cycle lie on a sphere in J-dimensional space having the least-squares solution at its center. This is not true more generally, however.

A Question: In both the consistent and inconsistent cases, the sequence $\{x^k\}$ of ART iterates is bounded, as Tanabe [202], and De Pierro and Iusem [109] have shown. The proof is easy in the consistent case. Is there an easy proof for the inconsistent case?

8.4 Regularization

In many remote-sensing applications the entries of the vector b are measured data and therefore noisy. At the same time, the matrix A describing the sensing process may be a simplification of the actual situation. Combined, the description $Ax = b$ may not be precisely true. In such cases, finding an exact solution, even if one exists, may not be desirable, and regularization is adopted. Imposing constraints on the vector x may also result in there not being a solution.

8.4.1 Norm-Constrained Least-Squares

To regularize the least-squares problem we can minimize not $\|b - Ax\|_2$, but, say,

$$f(x) = \|b - Ax\|_2^2 + \epsilon^2 \|x\|_2^2, \tag{8.37}$$

for some small $\epsilon > 0$. Now we are still trying to make $\|b - Ax\|_2$ small, but managing to keep $\|x\|_2$ from becoming too large in the process. This leads to a *norm-constrained least-squares* solution.

The minimizer of $f(x)$ is the unique solution $\hat{x}_\epsilon$ of the system

$$(A^\dagger A + \epsilon^2 I)x = A^\dagger b. \tag{8.38}$$

When I and J are large, we need ways to solve this system without having to deal with the matrix $A^\dagger A + \epsilon^2 I$. The Landweber method allows us to avoid $A^\dagger A$ in calculating the least-squares solution. Is there a similar method to use now? Yes, there is.

8.4.2 Regularizing Landweber's Algorithm

Our goal is to minimize the function $f(x)$ in Equation (8.37). Notice that this is equivalent to minimizing the function

$$F(x) = ||Bx - c||_2^2, \tag{8.39}$$

for

$$B = \begin{bmatrix} A \\ \epsilon I \end{bmatrix}, \tag{8.40}$$

and

$$c = \begin{bmatrix} b \\ 0 \end{bmatrix}, \tag{8.41}$$

where 0 denotes a column vector with all entries equal to zero. The Landweber iteration for the problem $Bx = c$ is

$$x^{k+1} = x^k + \alpha B^\dagger(c - Bx^k), \tag{8.42}$$

for $0 < \alpha < 2/\rho(B^\dagger B)$, where $\rho(B^\dagger B)$ is the largest eigenvalue, or the spectral radius, of $B^\dagger B$. Equation (8.42) can be written as

$$x^{k+1} = (1 - \alpha\epsilon^2)x^k + \alpha A^\dagger(b - Ax^k). \tag{8.43}$$

8.4.3 Regularizing the ART

We would like to get the regularized solution $\hat{x}_\epsilon$ by taking advantage of the faster convergence of the ART. Fortunately, there are ways to find $\hat{x}_\epsilon$, using only the matrix A and the ART algorithm. We discuss two methods for using ART to obtain regularized solutions of $Ax = b$. The first one is presented in [61], while the second one is due to Eggermont, Herman, and Lent [116].

In our first method we use ART to solve the system of equations given in matrix form by

$$\begin{bmatrix} A^\dagger & \epsilon I \end{bmatrix} \begin{bmatrix} u \\ v \end{bmatrix} = 0. \tag{8.44}$$

We begin with $u^0 = b$ and $v^0 = 0$. Then, the lower component of the limit vector is $v^\infty = -\epsilon\hat{x}_\epsilon$, while the upper limit is $u^\infty = b - A\hat{x}_\epsilon$.

The method of Eggermont *et al.* is similar. In their method we use ART to solve the system of equations given in matrix form by

$$\begin{bmatrix} A & \epsilon I \end{bmatrix} \begin{bmatrix} x \\ v \end{bmatrix} = b. \tag{8.45}$$

We begin at $x^0 = 0$ and $v^0 = 0$. Then, the limit vector has for its upper component $x^\infty = \hat{x}_\epsilon$, and $\epsilon v^\infty = b - A\hat{x}_\epsilon$.

However, we do not want to calculate $A^\dagger A + \epsilon^2 I$ when the matrix A is large. Fortunately, there are ways to find $\hat{x}_\epsilon$, using only the matrix A and the ART algorithm.

8.5 Avoiding the Limit Cycle

Generally, the greater the minimum value of $||Ax - b||_2^2$ the more the vectors of the LC are distinct from one another. There are several ways to avoid the LC in ART and to obtain a least-squares solution. One way is the *double ART* (DART) [54]:

8.5.1 Double ART (DART)

We know that any b can be written as $b = A\hat{x} + \hat{w}$, where $A^\dagger \hat{w} = 0$ and $\hat{x}$ is a minimizer of $||Ax - b||_2^2$. The vector $\hat{w}$ is the orthogonal projection of b onto the null space of the matrix transformation $A^\dagger$. Therefore, in Step 1 of DART we apply the ART algorithm to the consistent system of linear equations $A^\dagger w = 0$, beginning with $w^0 = b$. The limit is $w^\infty = \hat{w}$, the member of the null space of $A^\dagger$ closest to b. In Step 2, apply ART to the consistent system of linear equations $Ax = b - w^\infty = A\hat{x}$. The limit is then the minimizer of $||Ax - b||_2$ closest to x^0. Notice that we could also obtain the least-squares solution by applying ART to the system $A^\dagger y = A^\dagger b$, starting with $y^0 = 0$, to obtain the minimum-norm solution, which is $y = A\hat{x}$, and then applying ART to the system $Ax = y$.

8.5.2 Strongly Under-Relaxed ART

Another method for avoiding the LC is *strong under-relaxation*, due to Censor, Eggermont and Gordon [82]. Let $t > 0$. Replace the iterative step in ART with

$$x_j^{k+1} = x_j^k + t\overline{A_{ij}}(b_i - (Ax^k)_i). \tag{8.46}$$

In [82] it is shown that, as $t \to 0$, the vectors of the LC approach the geometric least-squares solution closest to x^0; a short proof is in [51]. Bertsekas [24] uses strong under-relaxation to obtain convergence of more general incremental methods.

8.5.3 Nonnegative Least-Squares

If there is no solution to a system of linear equations $Ax = b$, then we may seek a *least-squares* "solution", which is a minimizer of the function

$$f(x) = \frac{1}{2}\sum_{i=1}^{I}\left|(\sum_{m=1}^{J} A_{im}x_m) - b_i\right|^2 = ||Ax - b||^2. \tag{8.47}$$

The partial derivative of $f(x)$ with respect to the variable x_j is

$$\frac{\partial f}{\partial x_j}(x) = \sum_{i=1}^{I}\overline{A_{ij}}\Big((\sum_{m=1}^{J} A_{im}x_m) - b_i\Big). \tag{8.48}$$

Setting the gradient equal to zero, we find that to get a least-squares solution we must solve the system of equations

$$A^\dagger(Ax - b) = 0. \tag{8.49}$$

Now we consider what happens when the additional constraints $x_j \geq 0$ are imposed.

This problem becomes a convex programming problem. Let $\hat{x}$ be a solution of the nonnegatively constrained least-squares problem. According to the Karush-Kuhn-Tucker Theorem [126], for those values of j for which $\hat{x}_j$ is not zero the corresponding Lagrange multiplier is $\lambda_j^* = 0$ and $\frac{\partial f}{\partial x_j}(\hat{x}) = 0$. Therefore, if $\hat{x}_j \neq 0$,

$$0 = \sum_{i=1}^{I} \overline{A_{ij}} \Big(\big(\sum_{m=1}^{J} A_{im} \hat{x}_m \big) - b_i \Big). \tag{8.50}$$

Let Q be the I by K matrix obtained from A by deleting rows j for which $\hat{x}_j = 0$. Then we can write

$$Q^\dagger (A\hat{x} - b) = 0. \tag{8.51}$$

If Q has $K \geq I$ columns and has full rank, then $Q^\dagger$ is a one-to-one linear transformation, which implies that $A\hat{x} = b$. Therefore, when there is no nonnegative solution of $Ax = b$, and Q has full rank, which is the typical case, the Q must have fewer than I columns, which means that $\hat{x}$ has fewer than I non-zero entries.

This result has some practical implications in medical image reconstruction. In the hope of improving the resolution of the reconstructed image, we may be tempted to take J, the number of pixels, larger than I, the number of equations arising from photon counts or line integrals. Since the vector b consists of measured data, it is noisy and there may well not be a nonnegative solution of $Ax = b$. As a result, the image obtained by nonnegatively constrained least-squares will have at most $I - 1$ non-zero entries; many of the pixels will be zero and they will be scattered throughout the image, making it unusable for diagnosis. The reconstructed images resemble stars in a night sky, and, as a result, the theorem is sometimes described as the "night sky" theorem.

This "night sky" phenomenon is not restricted to least squares. The same thing happens with methods based on the Kullback-Leibler distance, such as MART, EMML and SMART.

8.6 Exercises

Ex. 8.1 *Prove Lemma 8.2.*

Ex. 8.2 *In $\mathbb{R}^2$ let C_1 be the closed lower half-space, and C_2 the epi-graph of the function $g : (0, +\infty) \to (0, +\infty)$ given by $g(t) = 1/t$. Show that the* *proximity function*

$$f(x) = \sum_{i=1}^{2} ||P_{C_i}x - x||_2^2, \tag{8.52}$$

has no minimizer.

Ex. 8.3 *Let* $f(x) = \frac{1}{2\gamma}\|x - P_C x\|_2^2$*, for some* $\gamma > 0$*. Show that*

$$x = \text{prox}_f(z) = (1 - \alpha)z + \alpha P_C z,$$

where $\alpha = \frac{1}{\gamma+1}$*. This tells us that relaxed orthogonal projections are also prox operators. Hint: Use Theorem 8.1 to show that* x *must satisfy the equation*

$$z = x + \frac{1}{\gamma}(x - P_C x).$$

Then show that $P_C z = P_C x$*.*

Chapter 9

Eigenvalue Bounds

As we discussed previously, a number of iterative methods that involve a matrix A place upper bounds on the step-length parameter in terms of the spectral radius of the matrix $A^\dagger A$. Since A is often quite large, finding decent estimates of $\rho(A^\dagger A)$ without having to calculate $A^\dagger A$ becomes important. In this chapter we obtain upper bounds on the spectral radius of positive-definite matrices and use these bounds in the selection of parameters in several iterative methods.

9.1 Introduction and Notation

In this section we review basic definitions and properties of eigenvalues and matrix norms.

9.1.1 Eigenvalues and Singular Values

Let A be a complex M by N matrix. It is often helpful to know how large the two-norm $\|Ax\|_2$ can be, relative to $\|x\|_2$; that is, we want to find a constant a so that

$$\|Ax\|_2/\|x\|_2 \leq a,$$

for all $x \neq 0$. We can reformulate the problem by asking how large $\|Au\|_2^2$ can be, subject to $\|u\|_2 = 1$. Using Lagrange multipliers, we discover that a unit vector u that maximizes $\|Au\|_2^2$ has the property that

$$A^\dagger Au = \lambda u,$$

for some constant λ. This leads to the more general problem discussed in this section.

Definition 9.1 *Given an N by N complex matrix S, we say that a complex number λ is an* eigenvalue *of S if there is a nonzero vector u with $Su = \lambda u$. The column vector u is then called an* eigenvector *of S associated with eigenvalue λ.*

Clearly, if u is an eigenvector of S, then so is cu, for any constant $c \neq 0$; therefore, it is common to choose eigenvectors to have norm equal to one.

Definition 9.2 *The* spectral radius *of S, denoted $\rho(S)$, is the largest value of $|\lambda|$, where λ denotes an eigenvalue of S.*

Definition 9.3 *A Hermitian matrix Q is said to be* nonnegative-definite *if all the eigenvalues of Q are nonnegative, and* positive-definite *if all the eigenvalues are positive.*

Definition 9.4 *Let A be an arbitrary matrix. A nonnegative number γ is a* singular value *for A if γ^2 is an eigenvalue of both $A^\dagger A$ and $AA^\dagger$.*

We present now a number of assertions without proof. Details can be found in almost any text on applied or computational linear algebra; see, for example, [174].

- For any square matrix S, the trace of S is the sum of its eigenvalues.
- For any square matrix S we have $\rho(S)^2 = \rho(S^2)$.
- The eigenvalues of a Hermitian matrix H are real.
- A Hermitian matrix Q is a nonnegative-definite matrix if and only if there is another matrix C, not necessarily square, such that $Q = C^\dagger C$.

9.1.2 Vector and Matrix Norms

We consider now the most common norms on the space $\mathbb{C}^J$. These notions apply equally to $\mathbb{R}^J$.

The 1-norm on $\mathbb{C}^J$ is defined by

$$\|x\|_1 = \sum_{j=1}^{J} |x_j|. \tag{9.1}$$

The ∞-norm on $\mathbb{C}^J$ is defined by

$$\|x\|_\infty = \max\{|x_j|\,|j = 1, ..., J\}. \tag{9.2}$$

For any $p \geq 1$, the p-norm is defined by

$$\|x\|_p = \Big(\sum_{j=1}^{J} |x_j|^p\Big)^{1/p}. \tag{9.3}$$

The 2-norm, also called the Euclidean norm, is the most commonly used norm on $\mathbb{C}^J$. It is the p-norm for $p = 2$ and is the one that comes from the inner product:

$$\|x\|_2 = \sqrt{\sum_{j=1}^{J} |x_j|^2} = \sqrt{\langle x, x\rangle} = \sqrt{x^\dagger x}. \tag{9.4}$$

Any matrix can be turned into a vector by vectorization. Therefore, we can define a norm for any matrix by simply vectorizing the matrix and taking a norm of the resulting vector; the 2-norm of the vectorized matrix is the *Frobenius norm* of the matrix itself. Such norms for matrices may not be compatible with the role of a matrix as representing a linear transformation. For that reason, we consider norms on matrices that are induced by the norms of the vectors on which the matrices operate.

Definition 9.5 *Let A be an M by N complex matrix. A norm on A, denoted $\|A\|$, is said to be* compatible *with given norms on $\mathbb{C}^N$ and $\mathbb{C}^M$ if $\|Ax\| \leq \|A\|\|x\|$, for every x in $\mathbb{C}^N$.*

One way to obtain a compatible norm for matrices is through the use of an induced matrix norm.

Definition 9.6 *Let $\|x\|$ be any norm on $\mathbb{C}^J$, not necessarily the Euclidean norm, $\|b\|$ any norm on $\mathbb{C}^I$, and A a rectangular I by J matrix. The* induced matrix norm *of A, simply denoted $\|A\|$, derived from these two vector norms, is the smallest positive constant c such that*

$$\|Ax\| \leq c\|x\|, \tag{9.5}$$

for all x in $\mathbb{C}^J$. This induced norm can be written as

$$\|A\| = \max_{x \neq 0}\{\|Ax\|/\|x\|\}. \tag{9.6}$$

When A is square we always assume that it is the same norm being used on x and Ax.

9.1.3 Some Examples of Induced Matrix Norms

If we choose the two vector norms carefully, then we can get an explicit description of $\|A\|$, but, in general, we cannot.

For example, let $\|x\| = \|x\|_1$ and $\|Ax\| = \|Ax\|_1$ be the 1-norms of the vectors x and Ax. Then the 1-norm of A, induced by the 1-norms of vectors in $\mathbb{C}^J$ and $\mathbb{C}^I$, is

$$\|A\|_1 = \max\,\{\sum_{i=1}^{I} |A_{ij}|\,, j = 1, 2, ..., J\}. \tag{9.7}$$

The infinity norm of the matrix A, induced by the infinity norms of vectors in $\mathbb{C}^J$ and $\mathbb{C}^I$, is

$$\|A\|_\infty = \max\,\{\sum_{j=1}^{J} |A_{ij}|\,, i = 1, 2, ..., I\}. \tag{9.8}$$

Consequently,

$$\|A^\dagger\|_1 = \|A\|_\infty,$$

and

$$\|A^\dagger\|_\infty = \|A\|_1.$$

We shall be particularly interested in the two-norm (or 2-norm) of a matrix A, denoted by $\|A\|_2$, which is the induced matrix norm derived from the Euclidean vector norms.

From the definition of the two-norm of A, we know that

$$\|A\|_2 = \max\{\|Ax\|_2/\|x\|_2\}, \tag{9.9}$$

with the maximum over all nonzero vectors x. Since

$$\|Ax\|_2^2 = x^\dagger A^\dagger Ax, \tag{9.10}$$

we have

$$\|A\|_2 = \sqrt{\max\{\frac{x^\dagger A^\dagger Ax}{x^\dagger x}\}}, \tag{9.11}$$

over all nonzero vectors x. The two-norm of a matrix A is

$$\|A\|_2 = \sqrt{\rho(A^\dagger A)}; \tag{9.12}$$

that is, the term inside the square-root in Equation (9.11) is the largest eigenvalue of the matrix $A^\dagger A$. Therefore, $\|A\|_2 = \|A^\dagger\|_2$ for any matrix A.

Let H be an Hermitian matrix. The two-norm of H is $\|H\|_2 = \rho(H)$. We can use this to show that, for any matrix A, we have the inequality

$$\|A\|_2^2 \leq \|A\|_1\|A\|_\infty. \tag{9.13}$$

The inequality (9.13) also follows, as a particular case, from the more general Theorem 9.5 concerning upper bounds for the singular values of a matrix A. If the rows of the matrix A are rescaled so that, for each i, we have $\sum_{j=1}^J |A_{ij}| \leq 1$, then no eigenvalue of $A^\dagger A$ is larger than the maximum number of non-zero entries in any column of A. Later in this chapter we shall see that the same conclusion holds if the rows of A are rescaled to have Euclidean length not greater than one.

If S is not Hermitian, then the two-norm of S cannot be calculated directly from the eigenvalues of S. Take, for example, the square, non-Hermitian matrix

$$S = \begin{bmatrix} i & 2 \\ 0 & i \end{bmatrix}, \tag{9.14}$$

having eigenvalues $\lambda = i$ and $\lambda = i$. The eigenvalues of the Hermitian matrix

$$S^\dagger S = \begin{bmatrix} 1 & -2i \\ 2i & 5 \end{bmatrix} \tag{9.15}$$

are $\lambda = 3 + 2\sqrt{2}$ and $\lambda = 3 - 2\sqrt{2}$. Therefore, the two-norm of S is

$$\|S\|_2 = \sqrt{3 + 2\sqrt{2}}. \tag{9.16}$$

9.2 Overview

We are concerned here with iterative methods for solving, at least approximately, the system of I linear equations in J unknowns symbolized by $Ax = b$. In the applications of interest to us, such as medical imaging, both I and J are quite large, making the use of iterative methods the only feasible approach. It is also typical of such applications that the matrix A is sparse, that is, has relatively few non-zero entries. Therefore, iterative methods that exploit this sparseness to accelerate convergence are of special interest to us.

Cimmino's method [98] is a *simultaneous* method, in which all the equations are used at each step. The current vector x^{k-1} is projected orthogonally onto each of the hyperplanes and these projections are averaged to obtain the next iterate x^k. The iterative step of Cimmino's method is

$$x_j^k = \frac{1}{I}\sum_{i=1}^{I}\left(x_j^{k-1} + \overline{A_{ij}}\left(\frac{b_i - (Ax^{k-1})_i}{\sum_{t=1}^{J}|A_{it}|^2}\right)\right),$$

which can also be written as

$$x_j^k = x_j^{k-1} + \sum_{i=1}^{I}\overline{A_{ij}}\left(\frac{b_i - (Ax^{k-1})_i}{I\sum_{t=1}^{J}|A_{it}|^2}\right). \tag{9.17}$$

Landweber's iterative scheme [154] with

$$x^k = x^{k-1} + B^\dagger(d - Bx^{k-1}), \tag{9.18}$$

converges to the least-squares solution of $Bx = d$ closest to x^0, provided that the largest singular value of B does not exceed one. If we let B be the matrix with entries

$$B_{ij} = A_{ij}/\sqrt{I\sum_{t=1}^{J}|A_{it}|^2},$$

and define

$$d_i = b_i/\sqrt{I\sum_{t=1}^{J}|A_{it}|^2},$$

then, since the trace of the matrix $BB^\dagger$ is one, convergence of Cimmino's method follows. However, using the trace in this way to estimate the largest singular value of a matrix usually results in an estimate that is far too large, particularly when A is large and sparse, and therefore in an iterative algorithm with unnecessarily small step sizes.

The appearance of the term

$$I\sum_{t=1}^{J}|A_{it}|^2$$

in the denominator of Cimmino's method suggested to Censor et al. [88] that, when A is sparse, this denominator might be replaced with

$$\sum_{t=1}^{J}s_t|A_{it}|^2,$$

where s_t denotes the number of non-zero entries in the tth column of A. The resulting iterative method is the *component-averaging* (CAV) iteration. Convergence of the CAV method was established by showing that no singular value of the matrix B exceeds one, where B has the entries

$$B_{ij} = A_{ij}/\sqrt{\sum_{t=1}^{J}s_t|A_{it}|^2}.$$

In [65] we extended this result, to show that no eigenvalue of $A^\dagger A$ exceeds the maximum of the numbers

$$p_i = \sum_{t=1}^{J}s_t|A_{it}|^2.$$

Convergence of CAV then follows, as does convergence of several other methods, including the ART, Landweber's method, the SART [1], the block-iterative CAV (BICAV) [89], the CARP1 method of Gordon and Gordon [135], a block-iterative variant of CARP1 obtained from the DROP method of Censor et al. [84], and the SIRT method [205].

For a positive integer N with $1 \leq N \leq I$, we let $B_1, ..., B_N$ be not necessarily disjoint subsets of the set $\{i = 1, ..., I\}$; the subsets B_n are called *blocks*. We then let A_n be the matrix and b^n the vector obtained from A and b, respectively, by removing all the rows except for those whose index i is in the set B_n. For each n, we let s_{nt} be the number of non-zero entries in the tth column of the matrix A_n, s_n the maximum of the s_{nt}, s the maximum of the s_t, and $L_n = \rho(A_n^\dagger A_n)$ be the spectral radius, or largest eigenvalue, of the matrix $A_n^\dagger A_n$, with $L = \rho(A^\dagger A)$. We denote by A_i the ith row of the matrix A, and by ν_i the length of A_i, so that

$$\nu_i^2 = \sum_{j=1}^{J}|A_{ij}|^2.$$

9.3 Cimmino's Algorithm

The ART seeks a solution of $Ax = b$ by projecting the current vector x^{k-1} orthogonally onto the next hyperplane $H(a^{i(k)}, b_{i(k)})$ to get x^k; here $i(k) = k(\mathrm{mod}\,)I$. In Cimmino's algorithm, we project the current vector x^{k-1} onto each of the hyperplanes and then average the result to get x^k. The algorithm begins at $k = 1$, with an arbitrary x^0; the iterative step is then

$$x^k = \frac{1}{I}\sum_{i=1}^{I} P_i x^{k-1}, \tag{9.19}$$

where P_i is the orthogonal projection onto $H(a^i, b_i)$. The iterative step can then be written as

$$x_j^k = x_j^{k-1} + \frac{1}{I}\sum_{i=1}^{I}\left(\frac{\overline{A_{ij}}(b_i - (Ax^{k-1})_i)}{\nu_i^2}\right). \tag{9.20}$$

As we saw in our discussion of the ART, when the system $Ax = b$ has no solutions, the ART does not converge to a single vector, but to a limit cycle. One advantage of many simultaneous algorithms, such as Cimmino's, is that they do converge to the least-squares solution in the inconsistent case.

When $\nu_i = 1$ for all i, Cimmino's algorithm has the form $x^{k+1} = Tx^k$, for the operator T given by

$$Tx = (I - \frac{1}{I}A^\dagger A)x + \frac{1}{I}A^\dagger b.$$

Experience with Cimmino's algorithm shows that it is slow to converge. In the next section we consider how we might accelerate the algorithm.

9.4 The Landweber Algorithms

For simplicity, we assume, in this section, that $\nu_i = 1$ for all i. The Landweber algorithm [154, 23], with the iterative step

$$x^k = x^{k-1} + \gamma A^\dagger(b - Ax^{k-1}), \tag{9.21}$$

converges to the least-squares solution closest to the starting vector x^0, provided that $0 < \gamma < 2/\lambda_{max}$, where λ_{max} is the largest eigenvalue of the nonnegative-definite matrix $A^\dagger A$. Loosely speaking, the larger γ is, the faster the convergence. However, precisely because A is large, calculating the matrix $A^\dagger A$, not to mention finding its largest eigenvalue, can be prohibitively expensive. The matrix A is said to be sparse if most of its entries are zero. Useful upper bounds for λ_{max} are then given by Theorems 9.1 and 9.6.

9.4.1 Finding the Optimum γ

The operator

$$Tx = x + \gamma A^\dagger(b - Ax) = (I - \gamma A^\dagger A)x + \gamma A^\dagger b$$

is affine linear and is av if and only if its linear part, the Hermitian matrix

$$B = I - \gamma A^\dagger A,$$

is av. To guarantee this we need $0 \leq \gamma < 2/\lambda_{max}$. Should we always try to take γ near its upper bound, or is there an optimum value of γ? To answer this question we consider the eigenvalues of B for various values of γ.

Lemma 9.1 *If $\gamma < 0$, then none of the eigenvalues of B is less than one.*

Lemma 9.2 *For*

$$0 \leq \gamma \leq \frac{2}{\lambda_{max} + \lambda_{min}}, \tag{9.22}$$

we have

$$\rho(B) = 1 - \gamma\lambda_{min}; \tag{9.23}$$

the smallest value of $\rho(B)$ occurs when

$$\gamma = \frac{2}{\lambda_{max} + \lambda_{min}}, \tag{9.24}$$

and equals

$$\frac{\lambda_{max} - \lambda_{min}}{\lambda_{max} + \lambda_{min}}. \tag{9.25}$$

Similarly, for

$$\gamma \geq \frac{2}{\lambda_{max} + \lambda_{min}}, \tag{9.26}$$

we have

$$\rho(B) = \gamma\lambda_{max} - 1; \tag{9.27}$$

the smallest value of $\rho(B)$ *occurs when*

$$\gamma = \frac{2}{\lambda_{max} + \lambda_{min}}, \tag{9.28}$$

and equals

$$\frac{\lambda_{max} - \lambda_{min}}{\lambda_{max} + \lambda_{min}}. \tag{9.29}$$

We see from this lemma that, if $0 \leq \gamma < 2/\lambda_{max}$, and $\lambda_{min} > 0$, then $\|B\|_2 = \rho(B) < 1$, so that B is a strict contraction. We minimize $\|B\|_2$ by taking

$$\gamma = \frac{2}{\lambda_{max} + \lambda_{min}}, \tag{9.30}$$

in which case we have

$$\|B\|_2 = \frac{\lambda_{max} - \lambda_{min}}{\lambda_{max} + \lambda_{min}} = \frac{c-1}{c+1}, \tag{9.31}$$

for $c = \lambda_{max}/\lambda_{min}$, the *condition number* of the positive-definite matrix $A^\dagger A$. The closer c is to one, the smaller the norm $\|B\|_2$, and the faster the convergence.

On the other hand, if $\lambda_{min} = 0$, then $\rho(B) = 1$ for all γ in the interval $(0, 2/\lambda_{max})$. The matrix B is still averaged, but it is no longer a strict contraction. For example, consider the orthogonal projection P_0 onto the hyperplane $H_0 = H(a, 0)$, where $\|a\|_2 = 1$. This operator can be written

$$P_0 = I - aa^\dagger. \tag{9.32}$$

The largest eigenvalue of $aa^\dagger$ is $\lambda_{max} = 1$; the remaining ones are zero. The relaxed projection operator

$$B = I - \gamma aa^\dagger \tag{9.33}$$

has $\rho(B) = 1 - \gamma > 1$, if $\gamma < 0$, and for $\gamma \geq 0$, we have $\rho(B) = 1$. The operator B is averaged, in fact, it is firmly nonexpansive, but it is not a strict contraction.

9.4.2 The Projected Landweber Algorithm

When we require a nonnegative approximate solution x for the real system $Ax = b$ we can use a modified version of the Landweber algorithm, called the projected Landweber algorithm [23], in this case having the iterative step

$$x^{k+1} = (x^k + \gamma A^T(b - Ax^k))_+, \tag{9.34}$$

where, for any real vector a, we denote by $(a)_+$ the nonnegative vector whose entries are those of a, for those that are nonnegative, and are zero otherwise. The projected Landweber algorithm converges to a vector that minimizes $\|Ax - b\|_2$ over all nonnegative vectors x, for the same values of γ.

The projected Landweber algorithm is actually more general. For any closed, nonempty convex set C in $\mathbb{C}^J$, define the iterative sequence

$$x^{k+1} = P_C(x^k + \gamma A^\dagger(b - Ax^k)). \tag{9.35}$$

This sequence converges to a minimizer of the function $\|Ax - b\|_2$ over all x in C, whenever such minimizers exist.

Both the Landweber and projected Landweber algorithms are special cases of the CQ algorithm [58], which, in turn, is a special case of the more general iterative fixed point algorithm, with convergence governed by the Krasnosel'skii-Mann-Opial Theorem 7.1.

9.5 Some Upper Bounds for L

For the iterative algorithms we shall consider here, having a good upper bound for the largest eigenvalue of the matrix $A^\dagger A$ is important. In the applications of interest, principally medical image processing, the matrix A is large; even calculating $A^\dagger A$, not to mention computing eigenvalues, is prohibitively expensive. In addition, the matrix A is typically sparse, but $A^\dagger A$ will not be, in general. In this section we present upper bounds for L that are particularly useful when A is sparse and do not require the calculation of $A^\dagger A$.

9.5.1 Earlier Work

Many of the concepts we study in computational linear algebra were added to the mathematical toolbox relatively recently, as this area blossomed with the growth of electronic computers. Based on my brief investigations into the history of matrix theory, I believe that the concept of a norm of a matrix was not widely used prior to about 1945. This was recently confirmed when I read the paper [136]; as pointed out there, the use of matrix norms became an important part of numerical linear algebra only after the publication of [207]. Prior to the late 1940's a number of papers were published that established upper bounds on $\rho(A)$, for general square matrix A. As we now can see, several of these results are immediate consequences of the fact that $\rho(A) \leq \|A\|$, for any induced matrix norm. We give two examples.

For a given N by N matrix A, let

$$C_n = \sum_{m=1}^{N} |A_{mn}|,$$

$$R_m = \sum_{n=1}^{N} |A_{mn}|,$$

and C and R the maxima of C_n and R_m, respectively. We now know that $C = \|A\|_1$, and $R = \|A\|_\infty$, but the earlier authors did not.

In 1930 Browne [33] proved the following theorem.

Theorem 9.1 (Browne) *Let λ be any eigenvalue of A. Then*

$$|\lambda| \leq \frac{1}{2}(C + R).$$

In 1944 Farnell [124] published the following theorems.

Theorem 9.2 (Farnell I) *For any eigenvalue λ of A we have*

$$|\lambda| \leq \sqrt{CR}.$$

Theorem 9.3 (Farnell II) *Let*

$$r_m = \sum_{n=1}^{N} |A_{mn}|^2,$$

and

$$c_m = \sum_{n=1}^{N} |A_{nm}|^2.$$

Then, for any eigenvalue λ of A, we have

$$|\lambda| \leq \sqrt{\sum_{m=1}^{N} \sqrt{r_m c_m}}.$$

In 1946 Brauer [30] proved the following theorem.

Theorem 9.4 (Brauer) *For any eigenvalue λ of A, we have*

$$|\lambda| \leq \min\{C, R\}.$$

Let A be an arbitrary rectangular complex matrix. Since the largest singular value of A is the square root of the maximum eigenvalue of the square matrix $S = A^\dagger A$, we could use the inequality

$$\rho(A^\dagger A) = \|A^\dagger A\|_2 \leq \|A^\dagger A\|,$$

for any induced matrix norm, to establish an upper bound for the singular values of A. However, that bound would be in terms of the entries of $A^\dagger A$, not of A itself. In what follows we obtain upper bounds on the singular values of A in terms of the entries of A itself.

We see from Exercise 9.3 that Farnell (I) does generalize to arbitrary rectangular matrices and singular values. Brauer's Theorem 9.4 may suggest that no singular value of a rectangular matrix A exceeds the minimum of $\|A\|_1$ and $\|A\|_\infty$, but this is not true. Consider the matrix A whose SVD is given by

$$A = \begin{bmatrix} 4 & 3 \\ 8 & 6 \\ 8 & 6 \end{bmatrix} = \begin{bmatrix} 1/3 & 2/3 & 2/3 \\ 2/3 & -2/3 & 1/3 \\ 2/3 & 1/3 & -2/3 \end{bmatrix} \begin{bmatrix} 15 & 0 \\ 0 & 0 \\ 0 & 0 \end{bmatrix} \begin{bmatrix} 4/5 & 3/5 \\ 3/5 & -4/5 \end{bmatrix}.$$

The largest singular value of A is 15, $\|A\|_1 = 20$, $\|A\|_\infty = 14$, and we do have

$$15 \leq \sqrt{(20)(14)},$$

but we do not have

$$15 \leq \min\{20, 14\} = 14.$$

9.5.2 Our Basic Eigenvalue Inequality

In [205] van der Sluis and van der Vorst show that certain rescaling of the matrix A results in none of the eigenvalues of $A^\dagger A$ exceeding one. A modification of their proof leads to upper bounds on the eigenvalues of the original $A^\dagger A$ ([65]). For any a in the interval $[0, 2]$ let

$$c_{aj} = c_{aj}(A) = \sum_{i=1}^{I} |A_{ij}|^a,$$

$$r_{ai} = r_{ai}(A) = \sum_{j=1}^{J} |A_{ij}|^{2-a},$$

and c_a and r_a the maxima of the c_{aj} and r_{ai}, respectively. We prove the following theorem.

Theorem 9.5 *For any a in the interval* $[0,2]$, *no eigenvalue of the matrix* $A^\dagger A$ *exceeds the maximum of*

$$\sum_{j=1}^{J} c_{aj}|A_{ij}|^{2-a},$$

over all i, *nor the maximum of*

$$\sum_{i=1}^{I} r_{ai}|A_{ij}|^{a},$$

over all j. *Therefore, no eigenvalue of* $A^\dagger A$ *exceeds* $c_a r_a$.

Proof: Let $A^\dagger Av = \lambda v$, and let $w = Av$. Then we have

$$\|A^\dagger w\|_2^2 = \lambda\|w\|_2^2.$$

Applying Cauchy's Inequality, we obtain

$$\Big|\sum_{i=1}^{I}\overline{A_{ij}}w_i\Big|^2 \le \Big(\sum_{i=1}^{I}|A_{ij}|^{a/2}|A_{ij}|^{1-a/2}|w_i|\Big)^2$$

$$\le \Big(\sum_{i=1}^{I}|A_{ij}|^{a}\Big)\Big(\sum_{i=1}^{I}|A_{ij}|^{2-a}|w_i|^2\Big).$$

Therefore,

$$\|A^\dagger w\|_2^2 \le \sum_{j=1}^{J}\Big(c_{aj}\big(\sum_{i=1}^{I}|A_{ij}|^{2-a}|w_i|^2\big)\Big) = \sum_{i=1}^{I}\Big(\sum_{j=1}^{J}c_{aj}|A_{ij}|^{2-a}\Big)|w_i|^2$$

$$\le \max_i\Big(\sum_{j=1}^{J}c_{aj}|A_{ij}|^{2-a}\Big)\|w\|^2.$$

The second assertion follows in similar fashion. ∎

As a corollary, we obtain the following eigenvalue inequality, which is central to our discussion.

Corollary 9.1 *For each* $i = 1, 2, ..., I$, *let*

$$p_i = \sum_{j=1}^{J} s_j|A_{ij}|^2,$$

and let p *be the maximum of the* p_i. *Then* $L \le p$.

Proof: Take $a = 0$. Then, using the convention that $0^0 = 0$, we have $c_{0j} = s_j$. ∎

Corollary 9.2 *([58]; [204], Th. 4.2) If $\sum_{j=1}^J |A_{ij}|^2 \leq 1$ for each i, then $L \leq s$.*

Proof: For all i we have

$$p_i = \sum_{j=1}^J s_j |A_{ij}|^2 \leq s \sum_{j=1}^J |A_{ij}|^2 \leq s.$$

Therefore,

$$L \leq p \leq s.$$

∎

Corollary 9.3 *Selecting $a = 1$, we have*

$$L = \|A\|_2^2 \leq \|A\|_1 \|A\|_\infty = c_1 r_1.$$

Therefore, the largest singular value of A does not exceed $\sqrt{\|A\|_1 \|A\|_\infty}$.

Corollary 9.4 *Selecting $a = 2$, we have*

$$L = \|A\|_2^2 \leq \|A\|_F^2,$$

where $\|A\|_F$ denotes the Frobenius norm of A.

Corollary 9.5 *Let G be the matrix with entries*

$$G_{ij} = A_{ij} \sqrt{\alpha_i} \sqrt{\beta_j},$$

where

$$\alpha_i \leq \Big(\sum_{j=1}^J s_j \beta_j |A_{ij}|^2 \Big)^{-1},$$

for all i. Then $\rho(G^\dagger G) \leq 1$.

Proof: We have

$$\sum_{j=1}^J s_j |G_{ij}|^2 = \alpha_i \sum_{j=1}^J s_j \beta_j |A_{ij}|^2 \leq 1,$$

for all i. The result follows from Corollary 9.1. ∎

Corollary 9.6 *If $\sum_{j=1}^J s_j |A_{ij}|^2 \leq 1$ for all i, then $L \leq 1$.*

Corollary 9.7 *If* $0 < \gamma_i \leq p_i^{-1}$ *for all* i*, then the matrix* B *with entries* $B_{ij} = \sqrt{\gamma_i} A_{ij}$ *has* $\rho(B^\dagger B) \leq 1$.

Proof: We have

$$\sum_{j=1}^{J} s_j |B_{ij}|^2 = \gamma_i \sum_{j=1}^{J} s_j |A_{ij}|^2 = \gamma_i p_i \leq 1.$$

Therefore, $\rho(B^\dagger B) \leq 1$, according to the theorem. ∎

Corollary 9.8 *If, for some* a *in the interval* $[0, 2]$*, we have*

$$\alpha_i \leq r_{ai}^{-1}, \tag{9.36}$$

for each i*, and*

$$\beta_j \leq c_{aj}^{-1}, \tag{9.37}$$

for each j*, then, for the matrix* G *with entries*

$$G_{ij} = A_{ij} \sqrt{\alpha_i} \sqrt{\beta_j},$$

no eigenvalue of $G^\dagger G$ *exceeds one.*

Proof: We calculate $c_{aj}(G)$ and $r_{ai}(G)$ and find that

$$c_{aj}(G) \leq \left(\max_i \alpha_i^{a/2} \right) \beta_j^{a/2} \sum_{i=1}^{I} |A_{ij}|^a = \left(\max_i \alpha_i^{a/2} \right) \beta_j^{a/2} c_{aj}(A),$$

and

$$r_{ai}(G) \leq \left(\max_j \beta_j^{1-a/2} \right) \alpha_i^{1-a/2} r_{ai}(A).$$

Therefore, applying the inequalities (9.36) and (9.37), we have

$$c_{aj}(G) r_{ai}(G) \leq 1,$$

for all i and j. Consequently, $\rho(G^\dagger G) \leq 1$. ∎

9.5.3 Another Upper Bound for L

The next theorem ([58]) provides another upper bound for L that is useful when A is sparse. As previously, for each i and j, we let $e_{ij} = 1$, if A_{ij} is not zero, and $e_{ij} = 0$, if $A_{ij} = 0$. Let $0 < \nu_i = \sqrt{\sum_{j=1}^{J} |A_{ij}|^2}$, $\sigma_j = \sum_{i=1}^{I} e_{ij} \nu_i^2$, and σ be the maximum of the σ_j.

Theorem 9.6 *([58]) No eigenvalue of $A^\dagger A$ exceeds σ.*

Proof: Let $A^\dagger Av = cv$, for some non-zero vector v and scalar c. With $w = Av$, we have

$$w^\dagger AA^\dagger w = cw^\dagger w.$$

Then

$$\Big|\sum_{i=1}^{I}\overline{A_{ij}}w_i\Big|^2 = \Big|\sum_{i=1}^{I}\overline{A_{ij}}e_{ij}\nu_i\frac{w_i}{\nu_i}\Big|^2 \leq \Big(\sum_{i=1}^{I}|A_{ij}|^2\frac{|w_i|^2}{\nu_i^2}\Big)\Big(\sum_{i=1}^{I}\nu_i^2 e_{ij}\Big)$$

$$= \Big(\sum_{i=1}^{I}|A_{ij}|^2\frac{|w_i|^2}{\nu_i^2}\Big)\sigma_j \leq \sigma\Big(\sum_{i=1}^{I}|A_{ij}|^2\frac{|w_i|^2}{\nu_i^2}\Big).$$

Therefore, we have

$$cw^\dagger w = w^\dagger AA^\dagger w = \sum_{j=1}^{J}\Big|\sum_{i=1}^{I}\overline{A_{ij}}w_i\Big|^2$$

$$\leq \sigma\sum_{j=1}^{J}\Big(\sum_{i=1}^{I}|A_{ij}|^2\frac{|w_i|^2}{\nu_i^2}\Big) = \sigma\sum_{i=1}^{I}|w_i|^2 = \sigma w^\dagger w.$$

We conclude that $c \leq \sigma$. ∎

Corollary 9.9 *Let the rows of A have Euclidean length one. Then no eigenvalue of $A^\dagger A$ exceeds the maximum number of non-zero entries in any column of A.*

Proof: We have $\nu_i^2 = \sum_{j=1}^{J}|A_{ij}|^2 = 1$, for each i, so that $\sigma_j = s_j$ is the number of non-zero entries in the jth column of A, and $\sigma = s$ is the maximum of the σ_j. ∎

Corollary 9.10 *Let ν be the maximum Euclidean length of any row of A and s the maximum number of non-zero entries in any column of A. Then $L \leq \nu^2 s$.*

When the rows of A have length one, it is easy to see that $L \leq I$, so the choice of $\gamma = \frac{1}{I}$ in the Landweber algorithm, which gives Cimmino's algorithm [98], is acceptable, although perhaps much too small.

The proof of Theorem 9.6 was suggested by results presented by Arnold Lent in informal discussions with Gabor Herman, Yair Censor, Rob Lewitt and me at MIPG in Philadelphia in the late 1990's.

9.6 Simultaneous Iterative Algorithms

In this section we apply the previous theorems to obtain convergence of several simultaneous iterative algorithms for linear systems.

9.6.1 The General Simultaneous Iterative Scheme

In this section we are concerned with simultaneous iterative algorithms having the following iterative step:

$$x_j^k = x_j^{k-1} + \lambda_k \sum_{i=1}^{I} \gamma_{ij} \overline{A_{ij}} (b_i - (Ax^{k-1})_i), \tag{9.38}$$

with $\lambda_k \in [\epsilon, 1]$ and the choices of the parameters γ_{ij} that guarantee convergence. Although we cannot prove convergence for this most general iterative scheme, we are able to prove the following theorems for the separable case of $\gamma_{ij} = \alpha_i \beta_j$.

Theorem 9.7 *If, for some a in the interval $[0, 2]$, we have*

$$\alpha_i \leq r_{ai}^{-1}, \tag{9.39}$$

for each i, and

$$\beta_j \leq c_{aj}^{-1}, \tag{9.40}$$

for each j, then the sequence $\{x^k\}$ given by Equation (9.38) converges to the minimizer of the proximity function

$$\sum_{i=1}^{I} \alpha_i |b_i - (Ax)_i|^2$$

for which

$$\sum_{j=1}^{J} \beta_j^{-1} |x_j - x_j^0|^2$$

is minimized.

Proof: For each i and j, let

$$G_{ij} = \sqrt{\alpha_i} \sqrt{\beta_j} A_{ij},$$

$$z_j = x_j / \sqrt{\beta_j},$$

and

$$d_i = \sqrt{\alpha_i} b_i.$$

Then $Ax = b$ if and only if $Gz = d$. From Corollary 9.8 we have that $\rho(G^\dagger G) \leq 1$. Convergence then follows from Theorem 8.3. ∎

Corollary 9.11 *Let* $\gamma_{ij} = \alpha_i \beta_j$*, for positive* α_i *and* β_j*. If*

$$\alpha_i \leq \Big(\sum_{j=1}^{J} s_j \beta_j |A_{ij}|^2\Big)^{-1}, \tag{9.41}$$

for each i*, then the sequence* $\{x^k\}$ *in (9.38) converges to the minimizer of the proximity function*

$$\sum_{i=1}^{I} \alpha_i |b_i - (Ax)_i|^2$$

for which

$$\sum_{j=1}^{J} \beta_j^{-1} |x_j - x_j^0|^2$$

is minimized.

Proof: We know from Corollary 9.5 that $\rho(G^\dagger G) \leq 1$. ∎

We now obtain convergence for several known algorithms as corollaries to the previous theorems.

9.6.2 The SIRT Algorithm

Corollary 9.12 *([205]) For some* a *in the interval* $[0, 2]$ *let* $\alpha_i = r_{ai}^{-1}$ *and* $\beta_j = c_{aj}^{-1}$*. Then the sequence* $\{x^k\}$ *in (9.38) converges to the minimizer of the proximity function*

$$\sum_{i=1}^{I} \alpha_i |b_i - (Ax)_i|^2$$

for which

$$\sum_{j=1}^{J} \beta_j^{-1} |x_j - x_j^0|^2$$

is minimized.

For the case of $a = 1$, the iterative step becomes

$$x_j^k = x_j^{k-1} + \sum_{i=1}^{I} \left(\frac{\overline{A_{ij}}(b_i - (Ax^{k-1})_i)}{(\sum_{t=1}^{J} |A_{it}|)(\sum_{m=1}^{I} |A_{mj}|)} \right),$$

which was considered in [139]. The SART algorithm [1] is a special case, in which it is assumed that $A_{ij} \geq 0$, for all i and j.

9.6.3 The CAV Algorithm

Corollary 9.13 *If* $\beta_j = 1$ *and* α_i *satisfies*

$$0 < \alpha_i \leq \Big(\sum_{j=1}^{J} s_j|A_{ij}|^2\Big)^{-1},$$

for each i*, then the algorithm with the iterative step*

$$x^k = x^{k-1} + \lambda_k \sum_{i=1}^{I} \alpha_i (b_i - (Ax^{k-1})_i) A_i^\dagger \tag{9.42}$$

converges to the minimizer of

$$\sum_{i=1}^{I} \alpha_i |b_i - (Ax^{k-1})_i|^2$$

for which $\|x - x^0\|$ *is minimized.*

When

$$\alpha_i = \Big(\sum_{j=1}^{J} s_j|A_{ij}|^2\Big)^{-1},$$

for each i, this is the relaxed *component-averaging* (CAV) method of Censor et al. [88].

9.6.4 The Landweber Algorithm

When $\beta_j = 1$ and $\alpha_i = \alpha$ for all i and j, we have the relaxed Landweber algorithm. The convergence condition in Equation (9.36) becomes

$$\alpha \leq \Big(\sum_{j=1}^{J} s_j|A_{ij}|^2\Big)^{-1} = p_i^{-1}$$

for all i, so $\alpha \leq p^{-1}$ suffices for convergence. Actually, the sequence $\{x^k\}$ converges to the minimizer of $\|Ax - b\|_2$ for which the distance $\|x - x^0\|_2$ is minimized, for any starting vector x^0, when $0 < \alpha < 1/L$. Easily obtained estimates of L are usually over-estimates, resulting in overly conservative choices of α. For example, if A is first normalized so that $\sum_{j=1}^{J} |A_{ij}|^2 = 1$ for each i, then the trace of $A^\dagger A$ equals I, which tells us that $L \leq I$. But this estimate, which is the one used in Cimmino's method [98], is far too large when A is sparse.

9.6.5 The Simultaneous DROP Algorithm

Corollary 9.14 *Let* $0 < w_i \leq 1$,

$$\alpha_i = w_i \nu_i^{-2} = w_i \Big(\sum_{j=1}^{J} |A_{ij}|^2 \Big)^{-1}$$

and $\beta_j = s_j^{-1}$, *for each* i *and* j. *Then the simultaneous algorithm with the iterative step*

$$x_j^k = x_j^{k-1} + \lambda_k \sum_{i=1}^{I} \left(\frac{w_i \overline{A_{ij}} (b_i - (Ax^{k-1})_i)}{s_j \nu_i^2} \right), \tag{9.43}$$

converges to the minimizer of the function

$$\sum_{i=1}^{I} \left| \frac{w_i (b_i - (Ax)_i)}{\nu_i} \right|^2$$

for which the function

$$\sum_{j=1}^{J} s_j |x_j - x_j^0|^2$$

is minimized.

For $w_i = 1$, this is the CARP1 algorithm of [135] (see also [88, 89]). The simultaneous DROP algorithm of [84] requires only that the weights w_i be positive, but dividing each w_i by their maximum, $\max_i\{w_i\}$, while multiplying each λ_k by the same maximum, gives weights in the interval $(0, 1]$. For convergence of their algorithm, we need to replace the condition $\lambda_k \leq 2 - \epsilon$ with $\lambda_k \leq \frac{2-\epsilon}{\max_i\{w_i\}}$.

The denominator in CAV is

$$\sum_{t=1}^{J} s_t |A_{it}|^2,$$

while that in CARP1 is

$$s_j \sum_{t=1}^{J} |A_{it}|^2.$$

It was reported in [135] that the two methods differed only slightly in the simulated cases studied.

9.7 Block-iterative Algorithms

The methods discussed in the previous section are *simultaneous*, that is, all the equations are employed at each step of the iteration. We turn now to *block-iterative methods*, which employ only some of the equations at each step. When the parameters are appropriately chosen, block-iterative methods can be significantly faster than simultaneous ones.

9.7.1 The Block-Iterative Landweber Algorithm

For a given set of blocks, the block-iterative Landweber algorithm has the following iterative step: with $n = k(\mathrm{mod}\, N)$,

$$x^k = x^{k-1} + \gamma_n A_n^\dagger (b^n - A_n x^{k-1}). \tag{9.44}$$

The sequence $\{x^k\}$ converges to the solution of $Ax = b$ that minimizes $\|x - x^0\|_2$, whenever the system $Ax = b$ has solutions, provided that the parameters γ_n satisfy the inequalities $0 < \gamma_n < 1/L_n$. This follows from Theorem 8.3 by replacing the matrices A_n with $\sqrt{\gamma_n} A_n$ and the vectors b^n with $\sqrt{\gamma_n} b^n$.

If the rows of the matrices A_n are normalized to have length one, then we know that $L_n \leq s_n$. Therefore, we can use parameters γ_n that satisfy

$$0 < \gamma_n \leq \left(s_n \sum_{j=1}^{J} |A_{ij}|^2 \right)^{-1}, \tag{9.45}$$

for each $i \in B_n$.

9.7.2 The BICAV Algorithm

We can extend the block-iterative Landweber algorithm as follows: let $n = k(\mathrm{mod}\, N)$ and

$$x^k = x^{k-1} + \lambda_k \sum_{i \in B_n} \gamma_i (b_i - (Ax^{k-1})_i) A_i^\dagger. \tag{9.46}$$

It follows from Theorem 9.1 that, in the consistent case, the sequence $\{x^k\}$ converges to the solution of $Ax = b$ that minimizes $\|x - x^0\|$, provided that, for each n and each $i \in B_n$, we have

$$\gamma_i \leq \left(\sum_{j=1}^{J} s_{nj} |A_{ij}|^2 \right)^{-1}.$$

The BICAV algorithm [89] uses

$$\gamma_i = \Big(\sum_{j=1}^{J} s_{nj} |A_{ij}|^2 \Big)^{-1}.$$

The iterative step of BICAV is

$$x^k = x^{k-1} + \lambda_k \sum_{i \in B_n} \left(\frac{b_i - (Ax^{k-1})_i}{\sum_{t=1}^{J} s_{nt} |A_{it}|^2} \right) A_i^{\dagger}. \tag{9.47}$$

9.7.3 A Block-Iterative CARP1

The obvious way to obtain a block-iterative version of CARP1 would be to replace the denominator term

$$s_j \sum_{t=1}^{J} |A_{it}|^2$$

with

$$s_{nj} \sum_{t=1}^{J} |A_{it}|^2.$$

However, this is problematic, since we cannot redefine the vector of unknowns using $z_j = x_j \sqrt{s_{nj}}$, since this varies with n. In [84], this issue is resolved by taking τ_j to be not less than the maximum of the s_{nj}, and using the denominator

$$\tau_j \sum_{t=1}^{J} |A_{it}|^2 = \tau_j \nu_i^2.$$

A similar device is used in [147] to obtain a convergent block-iterative version of SART. The iterative step of DROP is

$$x_j^k = x_j^{k-1} + \lambda_k \sum_{i \in B_n} \left(\overline{A_{ij}} \frac{(b_i - (Ax^{k-1})_i)}{\tau_j \nu_i^2} \right). \tag{9.48}$$

Convergence of the DROP (*diagonally-relaxed orthogonal projection*) iteration follows from their Theorem 11. We obtain convergence as a corollary of our previous results.

The change of variables is $z_j = x_j \sqrt{\tau_j}$, for each j. Using our eigenvalue bounds, it is easy to show that the matrices C_n with entries

$$(C_n)_{ij} = \left(\frac{A_{ij}}{\sqrt{\tau_j} \nu_i} \right),$$

for all $i \in B_n$ and all j, have $\rho(C_n^\dagger C_n) \leq 1$. The resulting iterative scheme, which is equivalent to Equation (9.48), then converges, whenever $Ax = b$ is consistent, to the solution minimizing the proximity function

$$\sum_{i=1}^{I} \left| \frac{b_i - (Ax)_i}{\nu_i} \right|^2$$

for which the function

$$\sum_{j=1}^{J} \tau_j |x_j - x_j^0|^2$$

is minimized.

9.7.4 Using Sparseness

Suppose, for the sake of illustration, that each column of A has s non-zero elements, for some $s < I$, and we let $r = s/I$. Suppose also that the number of members of B_n is $I_n = I/N$ for each n, and that N is not too large. Then s_n is approximately equal to $rI_n = s/N$. On the other hand, unless A_n has only zero entries, we know that $s_n \geq 1$. Therefore, it is no help to select N for which $s/N < 1$. For a given degree of sparseness s we need not select N greater than s. The more sparse the matrix A, the fewer blocks we need to gain the maximum advantage from the rescaling, and the more we can benefit from parallelization in the calculations at each step of the algorithm in Equation (8.32).

9.8 Exercises

Ex. 9.1 *Prove Theorems 9.1, 9.2, and 9.4 using properties of matrix norms. Can you also prove Theorem 9.3 this way?*

Ex. 9.2 *Prove Lemma 9.1.*

Ex. 9.3 *Let A be an arbitrary rectangular matrix. Prove that no singular value of A exceeds $\sqrt{\|A\|_1 \|A\|_\infty}$.*

Ex. 9.4 (Computer Problem) *Compare the speed of convergence of the ART and Cimmino algorithms.*

Ex. 9.5 (Computer Problem) *By generating sparse matrices of various sizes, test the accuracy of the estimates of the largest singular-value given above.*

Chapter 10

Jacobi and Gauss-Seidel Methods

In this chapter we examine two well known iterative algorithms for solving square systems of linear equations, the Jacobi method and the Gauss-Seidel method, in terms of averaged and paracontractive operators. Both these algorithms are easy to describe and to motivate. They both require not only that the system be square, that is, have the same number of unknowns as equations, but satisfy additional constraints needed for convergence.

Linear systems $Ax = b$ need not be square but can be associated with two square systems, $A^\dagger Ax = A^\dagger b$, the so-called *normal equations*, and $AA^\dagger z = b$, sometimes called the *Björck-Elfving equations* [107]. Both the Jacobi and the Gauss-Seidel algorithms can be modified to apply to any square system of linear equations, $Sz = h$. The resulting algorithms, the Jacobi overrelaxation (JOR) and successive overrelaxation (SOR) methods, involve the choice of a parameter. The JOR and SOR will converge for more general classes of matrices, provided that the parameter is appropriately chosen. Particular cases of the Jacobi and the Gauss-Seidel methods are equivalent to the Landweber algorithm and the ART, respectively.

When we say that an iterative method is convergent, or converges, under certain conditions, we mean that it converges for any consistent system of the appropriate type, and for any starting vector; any iterative method will converge if we begin at the right answer. We assume throughout this chapter that A is an I by J matrix.

10.1 The Jacobi and Gauss-Seidel Methods: An Example

Suppose we wish to solve the 3 by 3 system

$$S_{11}z_1 + S_{12}z_2 + S_{13}z_3 = h_1$$

$$S_{21}z_1 + S_{22}z_2 + S_{23}z_3 = h_2$$

$$S_{31}z_1 + S_{32}z_2 + S_{33}z_3 = h_3, \tag{10.1}$$

which we can rewrite as

$$z_1 = S_{11}^{-1}[h_1 - S_{12}z_2 - S_{13}z_3]$$

$$z_2 = S_{22}^{-1}[h_2 - S_{21}z_1 - S_{23}z_3]$$

$$z_3 = S_{33}^{-1}[h_3 - S_{31}z_1 - S_{32}z_2], \tag{10.2}$$

assuming that the diagonal terms S_{mm} are not zero. Let $z^0 = (z_1^0, z_2^0, z_3^0)^T$ be an initial guess for the solution. We then insert the entries of z^0 on the right sides and use the left sides to define the entries of the next guess z^1. This is one full cycle of *Jacobi's method.*

The Gauss-Seidel method is similar. Let $z^0 = (z_1^0, z_2^0, z_3^0)^T$ be an initial guess for the solution. We then insert z_2^0 and z_3^0 on the right side of the first equation, obtaining a new value z_1^1 on the left side. We then insert z_3^0 and z_1^1 on the right side of the second equation, obtaining a new value z_2^1 on the left. Finally, we insert z_1^1 and z_2^1 into the right side of the third equation, obtaining a new z_3^1 on the left side. This is one full cycle of the *Gauss-Seidel* (GS) method.

10.2 Splitting Methods

The Jacobi and the Gauss-Seidel methods are particular cases of a more general approach known as *splitting methods.* Splitting methods apply to square systems of linear equations. Let S be an arbitrary N by N square matrix, written as $S = M - K$. Then the linear system of equations $Sz = h$ is equivalent to $Mz = Kz + h$. If M is invertible, then we can also write $z = M^{-1}Kz + M^{-1}h$. This last equation suggests a class of iterative methods for solving $Sz = h$ known as *splitting methods*. The idea is to select a matrix M so that the equation

$$Mz^{k+1} = Kz^k + h \tag{10.3}$$

can be easily solved to get z^{k+1}; in the Jacobi method M is diagonal, and in the Gauss-Seidel method, M is triangular. Then we write

$$z^{k+1} = M^{-1}Kz^k + M^{-1}h. \tag{10.4}$$

From $K = M - S$, we can write Equation (10.4) as

$$z^{k+1} = z^k + M^{-1}(h - Sz^k). \tag{10.5}$$

Suppose that S is invertible and $\hat{z}$ is the unique solution of $Sz = h$. The error we make at the k-th step is $e^k = \hat{z} - z^k$, so that

$$e^{k+1} = M^{-1}Ke^k.$$

We want the error to decrease with each step, which means that we should seek M and K so that $||M^{-1}K|| < 1$. If S is not invertible and there are multiple solutions of $Sz = h$, then we do not want $M^{-1}K$ to be a strict contraction, but only av or pc. The operator T defined by

$$Tz = M^{-1}Kz + M^{-1}h = Bz + d \tag{10.6}$$

is an affine linear operator and will be a pc or av operator whenever $B = M^{-1}K$ is.

It follows from our previous discussion concerning linear av operators that, if $B = B^\dagger$ is Hermitian, then B is av if and only if

$$-1 < \lambda \leq 1, \tag{10.7}$$

for all (necessarily real) eigenvalues λ of B.

In general, though, the matrix $B = M^{-1}K$ will not be Hermitian, and deciding if such a non-Hermitian matrix is av is not a simple matter. We do know that, if B is av, so is $B^\dagger$; the matrix B is a convex combination of the identity and a nonexpansive matrix N, so $B^\dagger$ is a convex combination of the identity and $N^\dagger$, which is also nonexpansive, since $||N^\dagger|| = ||N|| \leq 1$. Consequently, the Hermitian matrix $Q = \frac{1}{2}(B + B^\dagger)$ is also av. Therefore, $I - Q = \frac{1}{2}(M^{-1}S + (M^{-1}S)^\dagger)$ is ism, and so is nonnegative definite. We have $-1 < \lambda \leq 1$, for any eigenvalue λ of Q.

Alternatively, we can use the EKN Theorem 7.3. According to that theorem, if B has a basis of eigenvectors, and $|\lambda| < 1$ for all eigenvalues λ of B that are not equal to one, then $\{z^k\}$ will converge to a solution of $Sz = h$, whenever solutions exist.

In what follows we shall write an arbitrary square matrix S as

$$S = L + D + U, \tag{10.8}$$

where L is the strictly lower triangular part of S, D the diagonal part, and U the strictly upper triangular part. When $S = H$ is Hermitian, we have

$$H = L + D + L^{\dagger}. \tag{10.9}$$

We list now several examples of iterative algorithms obtained by the splitting method. In the remainder of the chapter we discuss these methods in more detail.

10.3 Some Examples of Splitting Methods

As we shall now see, the Jacobi and Gauss-Seidel methods, as well as their overrelaxed versions, JOR and SOR, are splitting methods.

Jacobi's Method: Jacobi's method uses $M = D$ and $K = -L - U$, under the assumption that D is invertible. The matrix B is

$$B = M^{-1}K = -D^{-1}(L + U). \tag{10.10}$$

The Gauss-Seidel Method: The Gauss-Seidel (GS) method uses the splitting $M = D + L$, so that the matrix B is

$$B = I - (D + L)^{-1}S. \tag{10.11}$$

The Jacobi Overrelaxation Method (JOR): The JOR uses the splitting

$$M = \frac{1}{\omega}D \tag{10.12}$$

and

$$K = M - S = (\frac{1}{\omega} - 1)D - L - U. \tag{10.13}$$

The matrix B is

$$B = M^{-1}K = (I - \omega D^{-1}S). \tag{10.14}$$

The Successive Overrelaxation Method (SOR): The SOR uses the splitting $M = (\frac{1}{\omega}D + L)$, so that

$$B = M^{-1}K = (D + \omega L)^{-1}[(1 - \omega)D - \omega U] \tag{10.15}$$

or

$$B = I - \omega(D + \omega L)^{-1}S, \tag{10.16}$$

or

$$B = (I + \omega D^{-1}L)^{-1}[(1-\omega)I - \omega D^{-1}U]. \tag{10.17}$$

10.4 Jacobi's Algorithm and JOR

Let A be an invertible J by J matrix. For all x in $\mathbb{C}^J$, let$\|x\|_A = \|Ax\|$. For all J by J matrices B define

$$\|B\|_A = \|ABA^{-1}\|; \tag{10.18}$$

this is the induced matrix norm determined by the vector norm $\|x\|_A$. We have the following lemma [8].

Lemma 10.1 *Given B and $\epsilon > 0$, there is invertible A such that*

$$\|B\|_A \leq \rho(B) + \epsilon.$$

Corollary 10.1 *If $\rho(B) < 1$, then there is a vector norm for which the induced matrix norm of B is less than one.*

The matrix B in Equation (10.10) is not generally av and the Jacobi iterative scheme will not converge, in general. Additional conditions need to be imposed on S in order to guarantee convergence. One such condition is that S be strictly diagonally dominant. In that case, all the eigenvalues of $B = M^{-1}K$ can be shown to lie inside the unit circle of the complex plane, so that $\rho(B) < 1$.

It follows from Lemma 10.1 that B is sc with respect to some vector norm, and the Jacobi iteration converges. If, in addition, S is Hermitian, the eigenvalues of B are in the interval $(-1, 1)$, and so B is sc with respect to the Euclidean norm.

Alternatively, one has the *Jacobi overrelaxation* (JOR) method, which is essentially a special case of the Landweber algorithm and involves an arbitrary parameter.

For S an N by N matrix, Jacobi's method can be written as

$$z_m^{\text{new}} = S_{mm}^{-1}[h_m - \sum_{j \neq m} S_{mj} z_j^{\text{old}}], \tag{10.19}$$

for $m = 1, ..., N$. With D the invertible diagonal matrix with entries $D_{mm} = S_{mm}$ we can write one cycle of Jacobi's method as

$$z^{\mathrm{new}} = z^{\mathrm{old}} + D^{-1}(h - Sz^{\mathrm{old}}). \tag{10.20}$$

The *Jacobi overrelaxation* (JOR) method has the following full-cycle iterative step:

$$z^{\mathrm{new}} = z^{\mathrm{old}} + \omega D^{-1}(h - Sz^{\mathrm{old}}); \tag{10.21}$$

choosing $\omega = 1$ we get the Jacobi method. Convergence of the JOR iteration will depend, of course, on properties of S and on the choice of ω. When $S = Q$, where Q is Hermitian and nonnegative-definite, for example, $S = A^\dagger A$ or $S = AA^\dagger$, we can say more. Note that such Q can always be written in the form $Q = AA^\dagger$ or $Q = A^\dagger A$, for appropriately chosen A.

10.4.1 The JOR in the Nonnegative-Definite Case

When $S = Q$ is nonnegative-definite and the system $Qz = h$ is consistent the JOR converges to a solution for any $\omega \in (0, 2/\rho(D^{-1/2}QD^{-1/2}))$, where $\rho(Q)$ denotes the largest eigenvalue of the nonnegative-definite matrix Q. For nonnegative-definite Q, the convergence of the JOR method is implied by the KMO Theorem 7.1, since the JOR is equivalent to Landweber's algorithm in these cases. To see this, we rewrite Equation (10.21) as

$$v^{\mathrm{new}} = v^{\mathrm{old}} + \omega G^\dagger(f - Gv^{\mathrm{old}}),$$

where $v = D^{1/2}z$,

$$G^\dagger G = D^{-1/2}QD^{-1/2},$$

and

$$G^\dagger f = D^{-1/2}h.$$

The JOR method, as applied to $Qz = AA^\dagger z = b$, is equivalent to the Landweber iterative method for $Ax = b$.

Ex. 10.1 *Show that the system $AA^\dagger z = b$ has solutions whenever the system $Ax = b$ has solutions.*

Lemma 10.2 *If $\{z^k\}$ is the sequence obtained from the JOR, then the sequence $\{A^\dagger z^k\}$ is the sequence obtained by applying the Landweber algorithm to the system $D^{-1/2}Ax = D^{-1/2}b$, where D is the diagonal part of the matrix $Q = AA^\dagger$.*

If we select $\omega = 1/I$ we obtain the Cimmino method. Since the trace of the matrix $D^{-1/2}QD^{-1/2}$ equals I, which then is the sum of its eigenvalues, all of which are nonnegative, we know that $\omega = 1/I$ is less than two over the largest eigenvalue of the matrix $D^{-1/2}QD^{-1/2}$ and so this choice of ω is acceptable and the Cimmino algorithm converges whenever there are solutions of $Ax = b$. In fact, it can be shown that Cimmino's method converges to a least-squares approximate solution generally.

Similarly, the JOR method applied to the system $A^\dagger Ax = A^\dagger b$ is equivalent to the Landweber algorithm, applied to the system $Ax = b$.

Ex. 10.2 *Show that, if $\{z^k\}$ is the sequence obtained from the JOR, then the sequence $\{D^{1/2}z^k\}$ is the sequence obtained by applying the Landweber algorithm to the system $AD^{-1/2}x = b$, where D is the diagonal part of the matrix $S = A^\dagger A$.*

10.5 The Gauss-Seidel Algorithm and SOR

In general, the full-cycle iterative step of the Gauss-Seidel method is the following:

$$z^{\text{new}} = z^{\text{old}} + (D+L)^{-1}(h - Sz^{\text{old}}), \tag{10.22}$$

where $S = D + L + U$ is the decomposition of the square matrix S into its diagonal, lower triangular and upper triangular diagonal parts. The GS method does not converge without restrictions on the matrix S. As with the Jacobi method, strict diagonal dominance is a sufficient condition.

10.5.1 The Nonnegative-Definite Case

Now we consider the square system $Qz = h$, assuming that $Q = L + D + L^\dagger$ is Hermitian and nonnegative-definite, so that $x^\dagger Qx \geq 0$, for all x. It is easily shown that all the entries of D are nonnegative. We assume that all the diagonal entries of D are positive, so that $D + L$ is invertible. The Gauss-Seidel iterative step is $z^{k+1} = Tz^k$, where T is the affine linear operator given by $Tz = Bz + d$, for $B = -(D+L)^{-1}L^\dagger$ and $d = (D+L)^{-1}h$.

Proposition 10.1 *Let λ be an eigenvalue of B that is not equal to one. Then $|\lambda| < 1$.*

If B is diagonalizable, then, by Proposition 7.3, there is a norm with respect to which T is paracontractive, so, by the EKN Theorem 7.3, the GS iteration converges to a solution of $Qz = h$, whenever solutions exist.

Proof of Proposition (10.1): Let $Bv = \lambda v$, for v nonzero. Then $-Bv = (D+L)^{-1}L^{\dagger}v = -\lambda v$, so that

$$L^{\dagger}v = -\lambda(D+L)v, \tag{10.23}$$

and

$$Lv = -\overline{\lambda}(D+L)^{\dagger}v. \tag{10.24}$$

Therefore,

$$v^{\dagger}L^{\dagger}v = -\lambda v^{\dagger}(D+L)v. \tag{10.25}$$

Adding $v^{\dagger}(D+L)v$ to both sides, we get

$$v^{\dagger}Qv = (1-\lambda)v^{\dagger}(D+L)v. \tag{10.26}$$

Since the left side of the equation is real, so is the right side. Therefore

$$(1-\overline{\lambda})(D+L)^{\dagger}v = (1-\lambda)v^{\dagger}(D+L)v$$

$$= (1-\lambda)v^{\dagger}Dv + (1-\lambda)v^{\dagger}Lv$$

$$= (1-\lambda)v^{\dagger}Dv - (1-\lambda)\overline{\lambda}v^{\dagger}(D+L)^{\dagger}v. \tag{10.27}$$

So we have

$$[(1-\overline{\lambda}) + (1-\lambda)\overline{\lambda}]v^{\dagger}(D+L)^{\dagger}v = (1-\lambda)v^{\dagger}Dv, \tag{10.28}$$

or

$$(1-|\lambda|^2)v^{\dagger}(D+L)^{\dagger}v = (1-\lambda)v^{\dagger}Dv. \tag{10.29}$$

Multiplying by $(1-\overline{\lambda})$ on both sides, we get, on the left side,

$$(1-|\lambda|^2)v^{\dagger}(D+L)^{\dagger}v - (1-|\lambda|^2)\overline{\lambda}v^{\dagger}(D+L)^{\dagger}v, \tag{10.30}$$

which is equal to

$$(1-|\lambda|^2)v^{\dagger}(D+L)^{\dagger}v + (1-|\lambda|^2)v^{\dagger}Lv, \tag{10.31}$$

and, on the right side, we get

$$|1-\lambda|^2 v^{\dagger}Dv. \tag{10.32}$$

Consequently, we have

$$(1-|\lambda|^2)v^{\dagger}Qv = |1-\lambda|^2 v^{\dagger}Dv. \tag{10.33}$$

Since $v^{\dagger}Qv \geq 0$ and $v^{\dagger}Dv > 0$, it follows that $1-|\lambda|^2 \geq 0$. If $|\lambda| = 1$, then $|1-\lambda|^2 = 0$, so that $\lambda = 1$. This completes the proof. ∎

Note that $\lambda = 1$ if and only if $Qv = 0$. Therefore, if Q is invertible, the affine linear operator T is a strict contraction, and the GS iteration converges to the unique solution of $Qz = h$.

10.5.2 The GS Algorithm as ART

We show now that the GS algorithm, when applied to the system $Qz = AA^\dagger z = h$, is equivalent to the ART algorithm, applied to $Ax = b$. Let $AA^\dagger = Q = L + D + L^\dagger$.

It is convenient now to consider separately each sub-iteration step of the GS algorithm. For $m = 0, 1, ...$ and $i = m(\text{mod}\, I) + 1$, we denote by z^{m+1} the vector whose entries are

$$z_i^{m+1} = D_{ii}^{-1}\Big(h_i - (Qz^m)_i + Q_{ii}z_i^m\Big),$$

and $z_n^{m+1} = z_n^m$, for $n \neq i$. Therefore, we can write

$$z_i^{m+1} - z_i^m = D_{ii}^{-1}(b_i - (AA^\dagger z^m)_i).$$

Now let $x^m = A^\dagger z^m$ for each m. Then we have

$$x_j^{m+1} = (A^\dagger z^{m+1})_j = (A^\dagger z^m)_j + \overline{A_{ij}} D_{ii}^{-1}(b_i - (Ax^m)_i),$$

which is one step of the ART algorithm, applied to the system $Ax = b$. Note that

$$D_{ii} = \sum_{j=1}^{J} |A_{ij}|^2.$$

From this, we can conclude that if $\{z^k\}$ is the sequence produced by one step of the GS algorithm, applied to the system $AA^\dagger z = b$, then $\{x^k = A^\dagger z^k\}$ is the sequence produced by one full cycle of the ART algorithm, applied to the system $Ax = b$. Since we know that the ART algorithm converges whenever $Ax = b$ is consistent, we know now that the GS algorithm, applied to the system $AA^\dagger z = b$, converges whenever $Ax = b$ is consistent. So once again we have shown that when $S = Q$ is Hermitian and nonnegative definite, the GS method converges whenever there are solutions of $Qz = h$.

10.5.3 Successive Overrelaxation

The *successive overrelaxation* (SOR) method has the following full-cycle iterative step:

$$z^{\text{new}} = z^{\text{old}} + (\omega^{-1}D + L)^{-1}(h - Sz^{\text{old}}); \tag{10.34}$$

the choice of $\omega = 1$ gives the GS method. Convergence of the SOR iteration will depend, of course, on properties of S and on the choice of ω.

Using the form

$$B = (D + \omega L)^{-1}[(1 - \omega)D - \omega U] \tag{10.35}$$

we can show that

$$|\det(B)| = |1-\omega|^N. \tag{10.36}$$

From this and the fact that the determinant of B is the product of its eigenvalues, we conclude that $\rho(B) > 1$ if $\omega < 0$ or $\omega > 2$. When $S = Q$ is Hermitian and nonnegative-definite, we can say more.

10.5.4 The SOR for Nonnegative-Definite Q

When Q is nonnegative-definite and the system $Qz = h$ is consistent the SOR converges to a solution for any $\omega \in (0, 2)$. This follows from the convergence of the ART algorithm, since, for such Q, the SOR is equivalent to the ART, as we now show.

Now we write $Q = AA^\dagger$ and consider the SOR method applied to the Björck-Elfving equations $AA^\dagger z = b$. Rather than count a full cycle as one iteration, we now count as a single step the calculation of a single new entry. Therefore, for $k = 0, 1, ...$ the $k+1$-st step replaces the value z_i^k only, where $i = k(\mathrm{mod}\, I) + 1$. We have

$$z_i^{k+1} = (1-\omega)z_i^k + \omega D_{ii}^{-1}(b_i - \sum_{n=1}^{i-1} Q_{in}z_n^k - \sum_{n=i+1}^{I} Q_{in}z_n^k) \tag{10.37}$$

and $z_n^{k+1} = z_n^k$ for $n \neq i$. Now we calculate $x^{k+1} = A^\dagger z^{k+1}$:

$$x_j^{k+1} = x_j^k + \omega D_{ii}^{-1}\overline{A_{ij}}(b_i - (Ax^k)_i). \tag{10.38}$$

This is one step of the relaxed *algebraic reconstruction technique* (ART) applied to the original system of equations $Ax = b$. The relaxed ART converges to a solution, when solutions exist, for any $\omega \in (0, 2)$.

When $Ax = b$ is consistent, so is $AA^\dagger z = b$. We consider now the case in which $Q = AA^\dagger$ is invertible. Since the relaxed ART sequence $\{x^k = A^\dagger z^k\}$ converges to a solution x^∞, for any $\omega \in (0, 2)$, the sequence $\{AA^\dagger z^k\}$ converges to b. Since $Q = AA^\dagger$ is invertible, the SOR sequence $\{z^k\}$ then converges to $Q^{-1}b$.

10.6 Summary

We summarize the basic points of this chapter:

1. Splitting methods for solving $Sz = h$, for square matrix $S = M - K$, involve affine linear operators $Tx = Bx + d$, where $B = M^{-1}K$ and $d = M^{-1}h$;

2. T is av if and only if B is av; if B is Hermitian, then B is av if and only if $-1 < \lambda \leq 1$ for all eigenvalues λ of B;

3. If B is not Hermitian, but is diagonalizable, and $|\lambda| < 1$ unless $\lambda = 1$, then there is a norm for which T is pc;

4. If S is strictly diagonally dominant, then the Jacobi and Gauss-Seidel iterations converge;

5. When $S = Q$ is Hermitian and nonnegative definite, Q can be written as either $AA^\dagger$ or as $A^\dagger A$, for appropriately chosen A, and the JOR method is equivalent to Landweber's algorithm for either $D^{-1/2}Ax = D^{-1/2}b$ or $AD^{-1/2}x = b$;

6. When $S = Q$ is Hermitian and nonnegative definite, and we write $Q = AA^\dagger$, the SOR method is equivalent to the relaxed ART algorithm for $Ax = b$, and so converges whenever there are solutions, for $0 < \omega < 2$.

Chapter 11

The SMART and EMML Algorithms

We turn now to iterative algorithms involving nonnegative vectors and matrices. For such algorithms the two-norm will not play a major role. Instead, the Kullback-Leibler, or cross-entropy, distance will be our primary tool. Our main examples are the simultaneous multiplicative algebraic reconstruction technique (SMART), the expectation maximization maximum likelihood (EMML) algorithms, and various related methods.

11.1 The SMART Iteration

The SMART minimizes the function $f(x) = KL(Px, y)$, over nonnegative vectors x. Here y is a vector with positive entries, and P is a matrix with nonnegative entries, such that $s_j = \sum_{i=1}^{I} P_{ij} > 0$. Denote by $\mathcal{X}$ the set of all nonnegative x for which the vector Px has only positive entries.

Having found the vector x^{k-1}, the next vector in the SMART sequence is x^k, with entries given by

$$x_j^k = x_j^{k-1} \exp s_j^{-1} \Big(\sum_{i=1}^{I} P_{ij} \log(y_i/(Px^{k-1})_i) \Big). \tag{11.1}$$

11.2 The EMML Iteration

The EMML algorithm minimizes the function $f(x) = KL(y, Px)$, over nonnegative vectors x. Having found the vector x^{k-1}, the next vector in the EMML sequence is x^k, with entries given by

$$x_j^k = x_j^{k-1} s_j^{-1} \Big(\sum_{i=1}^{I} P_{ij} (y_i/(Px^{k-1})_i) \Big). \tag{11.2}$$

11.3 The EMML and the SMART as AM

In [47] the SMART was derived using the following alternating minimization (AM) approach.

For each $x \in \mathcal{X}$, let $r(x)$ and $q(x)$ be the I by J arrays with entries

$$r(x)_{ij} = x_j P_{ij} y_i/(Px)_i, \tag{11.3}$$

and

$$q(x)_{ij} = x_j P_{ij}. \tag{11.4}$$

In the iterative step of the SMART we get x^k by minimizing the function

$$KL(q(x), r(x^{k-1})) = \sum_{i=1}^{I} \sum_{j=1}^{J} KL(q(x)_{ij}, r(x^{k-1})_{ij})$$

over $x \geq 0$. Note that $KL(Px, y) = KL(q(x), r(x))$.

Similarly, the iterative step of the EMML is to minimize the function $KL(r(x^{k-1}), q(x))$ to get $x = x^k$. Note that $KL(y, Px) = KL(r(x), q(x))$. It follows from the identities established in [47] that the SMART can also be formulated as a particular case of the SUMMA.

11.4 The SMART as SUMMA

We show now that the SMART is a particular case of the SUMMA. Lemma 2.1 is helpful in that regard. For notational convenience, we assume, for the remainder of this section, that $s_j = 1$ for all j. From the identities established for the SMART in [47], we know that the iterative step of SMART can be expressed as follows: minimize the function

$$G_k(x) = KL(Px, y) + KL(x, x^{k-1}) - KL(Px, Px^{k-1}) \tag{11.5}$$

to get x^k. According to Lemma 2.1, the quantity

$$g_k(x) = KL(x, x^{k-1}) - KL(Px, Px^{k-1})$$

is nonnegative, since $s_j = 1$. The $g_k(x)$ are defined for all nonnegative x; that is, the set D is the closed nonnegative orthant in $\mathbb{R}^J$. Each x^k is a positive vector.

It was shown in [47] that

$$G_k(x) = G_k(x^k) + KL(x, x^k), \tag{11.6}$$

from which it follows immediately that the SMART is in the SUMMA class.

Because the SMART is a particular case of the SUMMA, we know that the sequence $\{f(x^k)\}$ is monotonically decreasing to $f(\hat{x})$. It was shown in [47] that if $y = Px$ has no nonnegative solution and the matrix P and every submatrix obtained from P by removing columns has full rank, then $\hat{x}$ is unique; in that case, the sequence $\{x^k\}$ converges to $\hat{x}$. As we shall see, the SMART sequence always converges to a nonnegative minimizer of $f(x)$. To establish this, we reformulate the SMART as a particular case of the PMA.

11.5 The SMART as PMA

We take $F(x)$ to be the function

$$F(x) = \sum_{j=1}^{J} x_j \log x_j. \tag{11.7}$$

Then

$$D_F(x, z) = KL(x, z). \tag{11.8}$$

For nonnegative x and z in $\mathcal{X}$, we have

$$D_f(x,z) = KL(Px,Pz). \tag{11.9}$$

Lemma 11.1 $D_F(x,z) \geq D_f(x,z)$.

Proof: We have

$$D_F(x,z) \geq \sum_{j=1}^{J} KL(x_j,z_j) \geq \sum_{j=1}^{J}\sum_{i=1}^{I} KL(P_{ij}x_j,P_{ij}z_j)$$
$$\geq \sum_{i=1}^{I} KL((Px)_i,(Pz)_i) = KL(Px,Pz). \tag{11.10}$$

∎

We let $h(x) = F(x) - f(x)$; then $D_h(x,z) \geq 0$ for nonnegative x and z in $\mathcal{X}$. The iterative step of the SMART is to minimize the function

$$f(x) + D_h(x,x^{k-1}). \tag{11.11}$$

So the SMART is a particular case of the PMA.

The function $h(x) = F(x) - f(x)$ is finite on D the nonnegative orthant of $\mathbb{R}^J$, and differentiable on the interior, so $C = D$ is closed in this example. Consequently, $\hat{x}$ is necessarily in D. From our earlier discussion of the PMA, we can conclude that the sequence $\{D_h(\hat{x},x^k)\}$ is decreasing and the sequence $\{D_f(\hat{x},x^k)\} \to 0$. Since the function $KL(\hat{x},\cdot)$ has bounded level sets, the sequence $\{x^k\}$ is bounded, and $f(x^*) = f(\hat{x})$, for every cluster point. Therefore, the sequence $\{D_h(x^*,x^k)\}$ is decreasing. Since a subsequence converges to zero, the entire sequence converges to zero. The convergence of $\{x^k\}$ to x^* follows from basic properties of the KL distance.

From the fact that $\{D_f(\hat{x},x^k)\} \to 0$, we conclude that $P\hat{x} = Px^*$. Equation (4.17) now tells us that the difference $D_h(\hat{x},x^{k-1}) - D_h(\hat{x},x^k)$ depends only on $P\hat{x}$, and not directly on $\hat{x}$. Therefore, the difference $D_h(\hat{x},x^0) - D_h(\hat{x},x^*)$ also depends only on $P\hat{x}$ and not directly on $\hat{x}$. Minimizing $D_h(\hat{x},x^0)$ over nonnegative minimizers $\hat{x}$ of $f(x)$ is therefore equivalent to minimizing $D_h(\hat{x},x^*)$ over the same vectors. But the solution to the latter problem is obviously $\hat{x} = x^*$. Thus we have shown that the limit of the SMART is the nonnegative minimizer of $KL(Px,y)$ for which the distance $KL(x,x^0)$ is minimized.

The following theorem summarizes the situation with regard to the SMART.

Theorem 11.1 *In the consistent case the SMART converges to the unique nonnegative solution of $y = Px$ for which the distance $\sum_{j=1}^{J} s_j KL(x_j,x_j^0)$ is minimized. In the inconsistent case it converges to the unique nonnegative minimizer of the distance $KL(Px,y)$ for which $\sum_{j=1}^{J} s_j KL(x_j,x_j^0)$ is* *minimized; if P and every matrix derived from P by deleting columns has full rank then there is a unique nonnegative minimizer of $KL(Px, y)$ and at most $I - 1$ of its entries are nonzero.*

11.6 Using KL Projections

For each $i = 1, 2, ..., I$, let H_i be the hyperplane

$$H_i = \{z | (Pz)_i = y_i\}. \tag{11.12}$$

The KL projection of a given positive x onto H_i is the z in H_i that minimizes the KL distance $KL(z, x)$. Generally, the KL projection onto H_i cannot be expressed in closed form. However, the z in H_i that minimizes the weighted KL distance

$$\sum_{j=1}^{J} P_{ij} KL(z_j, x_j) \tag{11.13}$$

is $T_i(x)$ given by

$$T_i(x)_j = x_j y_i / (Px)_i. \tag{11.14}$$

Both the SMART and the EMML can be described in terms of the T_i.

The iterative step of the SMART algorithm can be expressed as

$$x_j^k = \prod_{i=1}^{I} (T_i(x^{k-1})_j)^{P_{ij}}. \tag{11.15}$$

We see that x_j^k is a weighted geometric mean of the terms $T_i(x^{k-1})_j$.

The iterative step of the EMML algorithm can be expressed as

$$x_j^k = \sum_{i=1}^{I} P_{ij} T_i(x^{k-1})_j. \tag{11.16}$$

We see that x_j^k is a weighted arithmetic mean of the terms $T_i(x^{k-1})_j$, using the same weights as in the case of SMART.

11.7 The MART and EMART Algorithms

The MART algorithm has the iterative step

$$x_j^k = x_j^{k-1}(y_i/(Px^{k-1})_i)^{P_{ij}m_i^{-1}}, \tag{11.17}$$

where $i = (k-1)(\mathrm{mod}\, I) + 1$ and

$$m_i = \max\{P_{ij} | j = 1, 2, ..., J\}. \tag{11.18}$$

When there are nonnegative solutions of the system $y = Px$, the sequence $\{x^k\}$ converges to the solution x that minimizes $KL(x, x^0)$ [50, 51, 52]. We can express the MART in terms of the weighted KL projections $T_i(x^{k-1})$;

$$x_j^k = (x_j^{k-1})^{1-P_{ij}m_i^{-1}}(T_i(x^{k-1})_j)^{P_{ij}m_i^{-1}}. \tag{11.19}$$

We see then that the iterative step of the MART is a relaxed weighted KL projection onto H_i, and a weighted geometric mean of the current x_j^k and $T_i(x^{k-1})_j$. The expression for the MART in Equation (11.19) suggests a somewhat simpler iterative algorithm involving a weighted arithmetic mean of the current x_j^{k-1} and $T_i(x^{k-1})_j$; this is the EMART algorithm.

The iterative step of the EMART algorithm is

$$x_j^k = (1 - P_{ij}m_i^{-1})x_j^{k-1} + P_{ij}m_i^{-1}T_i(x^{k-1})_j. \tag{11.20}$$

Whenever the system $y = Px$ has nonnegative solutions, the EMART sequence $\{x^k\}$ converges to a nonnegative solution, but nothing further is known about this solution. One advantage that the EMART has over the MART is the substitution of multiplication for exponentiation.

Block-iterative versions of SMART and EMML have also been investigated; see [50, 51, 52] and the references therein.

11.8 Extensions of MART and EMART

As we have seen, the iterative steps of the MART and the EMART are relaxed weighted KL projections onto the hyperplane H_i, resulting in vectors that are not within H_i. This suggests variants of MART and EMART in which, at the end of each iterative step, a further weighted KL projection onto H_i is performed. In other words, for MART and EMART the new vector would be $T_i(x^k)$, instead of x^k as given by Equations (11.17) and (11.20), respectively. Research into the properties of these new algorithms is ongoing.

11.9 Convergence of the SMART and EMML

In this section we prove convergence of the SMART and EMML algorithms through a series of exercises. For both algorithms we begin with an arbitrary positive vector x^0. The iterative step for the EMML method is

$$x_j^k = (x^{k-1})'_j = x_j^{k-1} \sum_{i=1}^{I} P_{ij} \frac{y_i}{(Px^{k-1})_i}. \tag{11.21}$$

The iterative step for the SMART is

$$x_j^m = (x^{m-1})''_j = x_j^{m-1} \exp\Big(\sum_{i=1}^{I} P_{ij} \log \frac{y_i}{(Px^{m-1})_i} \Big). \tag{11.22}$$

Note that, to avoid confusion, we use k for the iteration number of the EMML and m for the SMART.

11.9.1 Pythagorean Identities for the KL Distance

The SMART and EMML iterative algorithms are best derived using the principle of *alternating minimization*, according to which the distances $KL(r(x), q(z))$ and $KL(q(x), r(z))$ are minimized, first with respect to the variable x and then with respect to the variable z. Although the KL distance is not Euclidean, and, in particular, not even symmetric, there are analogues of Pythagoras' theorem that play important roles in the convergence proofs.

Ex. 11.1 *Establish the following Pythagorean identities:*

$$KL(r(x), q(z)) = KL(r(z), q(z)) + KL(r(x), r(z)); \tag{11.23}$$

$$KL(r(x), q(z)) = KL(r(x), q(x')) + KL(x', z), \tag{11.24}$$

for

$$x'_j = x_j \sum_{i=1}^{I} P_{ij} \frac{y_i}{(Px)_i}; \tag{11.25}$$

$$KL(q(x), r(z)) = KL(q(x), r(x)) + KL(x, z) - KL(Px, Pz); \tag{11.26}$$

$$KL(q(x), r(z)) = KL(q(z''), r(z)) + KL(x, z''), \tag{11.27}$$

for

$$z_j'' = z_j \exp(\sum_{i=1}^{I} P_{ij} \log \frac{y_i}{(Pz)_i}). \tag{11.28}$$

Note that it follows from Equation (2.13) that $KL(x,z) - KL(Px, Pz) \geq 0$.

11.9.2 Convergence Proofs

We shall prove convergence of the SMART and EMML algorithms through a series of exercises.

Ex. 11.2 *Show that, for* $\{x^k\}$ *given by Equation (11.21),* $\{KL(y, Px^k)\}$ *is decreasing and* $\{KL(x^{k+1}, x^k)\} \to 0$*. Show that, for* $\{x^m\}$ *given by Equation (11.22),* $\{KL(Px^m, y)\}$ *is decreasing and* $\{KL(x^m, x^{m+1})\} \to 0$*. Hint: Use* $KL(r(x), q(x)) = KL(y, Px)$*,* $KL(q(x), r(x)) = KL(Px, y)$*, and the Pythagorean identities.*

Ex. 11.3 *Show that the EMML sequence* $\{x^k\}$ *is bounded by showing that*

$$\sum_{j=1}^{J} x_j^{k+1} = \sum_{i=1}^{I} y_i.$$

Show that the SMART sequence $\{x^m\}$ *is bounded by showing that*

$$\sum_{j=1}^{J} x_j^{m+1} \leq \sum_{i=1}^{I} y_i.$$

Ex. 11.4 *Show that* $(x^*)' = x^*$ *for any cluster point* x^* *of the EMML sequence* $\{x^k\}$ *and that* $(x^*)'' = x^*$ *for any cluster point* x^* *of the SMART sequence* $\{x^m\}$*. Hint: Use* $\{KL(x^{k+1}, x^k)\} \to 0$ *and* $\{KL(x^m, x^{m+1})\} \to 0$.

Ex. 11.5 *Let* $\hat{x}$ *and* $\tilde{x}$ *minimize* $KL(y, Px)$ *and* $KL(Px, y)$*, respectively, over all* $x \geq 0$*. Then,* $(\hat{x})' = \hat{x}$ *and* $(\tilde{x})'' = \tilde{x}$*. Hint: Apply Pythagorean identities to* $KL(r(\hat{x}), q(\hat{x}))$ *and* $KL(q(\tilde{x}), r(\tilde{x}))$.

Note that, because of convexity properties of the KL distance, even if the minimizers $\hat{x}$ and $\tilde{x}$ are not unique, the vectors $P\hat{x}$ and $P\tilde{x}$ are unique.

Ex. 11.6 *For the EMML sequence $\{x^k\}$ with cluster point x^* and $\hat{x}$ as defined previously, we have the double inequality*

$$KL(\hat{x}, x^k) \geq KL(r(\hat{x}), r(x^k)) \geq KL(\hat{x}, x^{k+1}), \tag{11.29}$$

from which we conclude that the sequence $\{KL(\hat{x}, x^k)\}$ is decreasing and $KL(\hat{x}, x^*) < +\infty$. Hint: For the first inequality calculate $KL(r(\hat{x}), q(x^k))$ in two ways. For the second one, use $(x)'_j = \sum_{i=1}^I r(x)_{ij}$ and Lemma 2.1.

Ex. 11.7 *Show that, for the SMART sequence $\{x^m\}$ with cluster point x^* and $\tilde{x}$ as defined previously, we have*

$$KL(\tilde{x}, x^m) - KL(\tilde{x}, x^{m+1}) = KL(Px^{m+1}, y) - KL(P\tilde{x}, y) +$$

$$KL(P\tilde{x}, Px^m) + KL(x^{m+1}, x^m) - KL(Px^{m+1}, Px^m), \tag{11.30}$$

and so $KL(P\tilde{x}, Px^*) = 0$, the sequence $\{KL(\tilde{x}, x^m)\}$ is decreasing and $KL(\tilde{x}, x^*) < +\infty$. Hint: Expand $KL(q(\tilde{x}), r(x^m))$ using the Pythagorean identities.

Ex. 11.8 *For x^* a cluster point of the EMML sequence $\{x^k\}$ we have $KL(y, Px^*) = KL(y, P\hat{x})$. Therefore, x^* is a nonnegative minimizer of $KL(y, Px)$. Consequently, the sequence $\{KL(x^*, x^k)\}$ converges to zero, and so $\{x^k\} \to x^*$. Hint: Use the double inequality of Equation (11.29) and $KL(r(\hat{x}), q(x^*))$.*

Ex. 11.9 *For x^* a cluster point of the SMART sequence $\{x^m\}$ we have $KL(Px^*, y) = KL(P\tilde{x}, y)$. Therefore, x^* is a nonnegative minimizer of $KL(Px, y)$. Consequently, the sequence $\{KL(x^*, x^m)\}$ converges to zero, and so $\{x^m\} \to x^*$. Moreover,*

$$KL(\tilde{x}, x^0) \geq KL(x^*, x^0)$$

for all $\tilde{x}$ as before. Hints: Use Exercise 11.7. For the final assertion use the fact that the difference $KL(\tilde{x}, x^m) - KL(\tilde{x}, x^{m+1})$ is independent of the choice of $\tilde{x}$, since it depends only on $Px^* = P\tilde{x}$. Now sum over the index m.

11.10 Regularization

The "night sky" phenomenon that occurs in nonnegatively constrained least-squares also happens with methods based on the Kullback-Leibler distance, such as MART, EMML and SMART, requiring some sort of regularization.

11.10.1 The "Night-Sky" Problem

As we saw previously, the sequence $\{x^k\}$ generated by the EMML iterative step in Equation (11.2) converges to a nonnegative minimizer $\hat{x}$ of $f(x) = KL(y, Px)$, and we have

$$\hat{x}_j = \hat{x}_j s_j^{-1} \sum_{i=1}^{I} P_{ij} \frac{y_i}{(P\hat{x})_i}, \tag{11.31}$$

for all j. We consider what happens when there is no nonnegative solution of the system $y = Px$.

For those values of j for which $\hat{x}_j > 0$, we have

$$s_j = \sum_{i=1}^{I} P_{ij} = \sum_{i=1}^{I} P_{ij} \frac{y_i}{(P\hat{x})_i}. \tag{11.32}$$

Now let Q be the I by K matrix obtained from P by deleting rows j for which $\hat{x}_j = 0$. If Q has full rank and $K \geq I$, then Q^T is one-to-one, so that $1 = \frac{y_i}{(P\hat{x})_i}$ for all i, or $y = P\hat{x}$. But we are assuming that there is no nonnegative solution of $y = Px$. Consequently, we must have $K < I$ and $I - K$ of the entries of $\hat{x}$ are zero.

11.11 Modifying the KL distance

The SMART, EMML and their block-iterative versions are based on the Kullback-Leibler distance between nonnegative vectors and require that the solution sought be a nonnegative vector. To impose more general constraints on the entries of x we derive algorithms based on shifted KL distances, also called Fermi-Dirac generalized entropies.

For a fixed real vector u, the shifted KL distance $KL(x - u, z - u)$ is defined for vectors x and z having $x_j \geq u_j$ and $z_j \geq u_j$. Similarly, the shifted distance $KL(v-x, v-z)$ applies only to those vectors x and z for which $x_j \leq v_j$ and $z_j \leq v_j$. For $u_j \leq v_j$, the combined distance

$$KL(x-u, z-u) + KL(v-x, v-z)$$

is restricted to those x and z whose entries x_j and z_j lie in the interval $[u_j, v_j]$. Our objective is to mimic the derivation of the SMART, EMML and RBI methods, replacing KL distances with shifted KL distances, to obtain algorithms that enforce the constraints $u_j \leq x_j \leq v_j$, for each j. The algorithms that result are the ABMART and ABEMML block-iterative methods. These algorithms were originally presented in [53], in which the vectors u and v were called a and b, hence the names of the algorithms. As previously, we shall assume that the entries of the matrix P are nonnegative. We shall denote by B_n, $n = 1, ..., N$ a partition of the index set $\{i = 1, ..., I\}$ into blocks. For $k = 0, 1, ...$ let $n(k) = k(\text{mod}\, N) + 1$.

The projected Landweber algorithm can also be used to impose the restrictions $u_j \leq x_j \leq v_j$; however, the projection step in that algorithm is implemented by clipping, or setting equal to u_j or v_j values of x_j that would otherwise fall outside the desired range. The result is that the values u_j and v_j can occur more frequently than may be desired. One advantage of the AB methods is that the values u_j and v_j represent barriers that can only be reached in the limit and are never taken on at any step of the iteration.

11.12 The ABMART Algorithm

We assume that $(Pu)_i \leq y_i \leq (Pv)_i$ and seek a solution of $Px = y$ with $u_j \leq x_j \leq v_j$, for each j. The algorithm begins with an initial vector x^0 satisfying $u_j \leq x_j^0 \leq v_j$, for each j. Having calculated x^k, we take

$$x_j^{k+1} = \alpha_j^k v_j + (1 - \alpha_j^k) u_j, \tag{11.33}$$

with $n = n(k)$,

$$\alpha_j^k = \frac{c_j^k \prod^n (d_i^k)^{P_{ij}}}{1 + c_j^k \prod^n (d_i^k)^{P_{ij}}}, \tag{11.34}$$

$$c_j^k = \frac{(x_j^k - u_j)}{(v_j - x_j^k)}, \tag{11.35}$$

and

$$d_j^k = \frac{(y_i - (Pu)_i)((Pv)_i - (Px^k)_i)}{((Pv)_i - y_i)((Px^k)_i - (Pu)_i)}, \tag{11.36}$$

where $\prod^n$ denotes the product over those indices i in $B_{n(k)}$. Notice that, at each step of the iteration, x_j^k is a convex combination of the endpoints u_j and v_j, so that x_j^k lies in the interval $[u_j, v_j]$.

We have the following theorem concerning the convergence of the ABMART algorithm:

Theorem 11.2 *If there is a solution of the system $Px = y$ that satisfies the constraints $u_j \leq x_j \leq v_j$ for each j, then, for any N and any choice of the blocks B_n, the ABMART sequence converges to that constrained solution of $Px = y$ for which the Fermi-Dirac generalized entropic distance from x to x^0,*

$$KL(x - u, x^0 - u) + KL(v - x, v - x^0),$$

is minimized. If there is no constrained solution of $Px = y$, then, for $N = 1$, the ABMART sequence converges to the minimizer of

$$KL(Px - Pu, y - Pu) + KL(Pv - Px, Pv - y)$$

for which

$$KL(x - u, x^0 - u) + KL(v - x, v - x^0)$$

is minimized.

The proof is similar to that for RBI-SMART and is found in [53].

11.13 The ABEMML Algorithm

We make the same assumptions as in the previous section. The iterative step of the ABEMML algorithm is

$$x_j^{k+1} = \alpha_j^k v_j + (1 - \alpha_j^k) u_j, \tag{11.37}$$

where

$$\alpha_j^k = \gamma_j^k / d_j^k, \tag{11.38}$$

$$\gamma_j^k = (x_j^k - u_j) e_j^k, \tag{11.39}$$

$$\beta_j^k = (v_j - x_j^k) f_j^k, \tag{11.40}$$

$$d_j^k = \gamma_j^k + \beta_j^k, \tag{11.41}$$

$$e_j^k = \left(1 - \sum_{i \in B_n} P_{ij}\right) + \sum_{i \in B_n} P_{ij} \left(\frac{y_i - (Pu)_i}{(Px^k)_i - (Pu)_i}\right), \tag{11.42}$$

and

$$f_j^k = \left(1 - \sum_{i \in B_n} P_{ij}\right) + \sum_{i \in B_n} P_{ij} \left(\frac{(Pv)_i - y_i}{(Pv)_i - (Px^k)_i}\right). \tag{11.43}$$

We have the following theorem concerning the convergence of the ABE-MML algorithm:

Theorem 11.3 *If there is a solution of the system $Px = y$ that satisfies the constraints $u_j \leq x_j \leq v_j$ for each j, then, for any N and any choice of the blocks B_n, the ABEMML sequence converges to such a constrained solution of $Px = y$. If there is no constrained solution of $Px = y$, then, for $N = 1$, the ABEMML sequence converges to a constrained minimizer of*

$$KL(y - Pu, Px - Pu) + KL(Pv - y, Pv - Px).$$

The proof is similar to that for RBI-EMML and is to be found in [53]. In contrast to the ABMART theorem, this is all we can say about the limits of the ABEMML sequences.

Open Question: How does the limit of the ABEMML iterative sequence depend, in the consistent case, on the choice of blocks, and, in general, on the choice of x^0?

Chapter 12

Alternating Minimization

12.1 Alternating Minimization

As we have seen, the SMART is best derived as an alternating minimization (AM) algorithm. The main reference for alternating minimization is the paper [105] of Csiszár and Tusnády. As the authors of [206] remark, the geometric argument in [105] is "deep, though hard to follow". As we shall see, all AM methods for which the five-point property of [105] holds fall into the SUMMA class (see [66]).

The alternating minimization (AM) approach provides a useful framework for the derivation of iterative optimization algorithms. In this section we discuss the five-point property of [105] and use it to obtain a somewhat simpler proof of convergence for their AM algorithm. We then show that all AM algorithms with the five-point property are in the SUMMA class.

12.1.1 The AM Framework

Suppose that P and Q are arbitrary non-empty sets and the function $\Theta(p,q)$ satisfies $-\infty < \Theta(p,q) \leq +\infty$, for each $p \in P$ and $q \in Q$. We assume that, for each $p \in P$, there is $q \in Q$ with $\Theta(p,q) < +\infty$. Therefore, $b = \inf_{p\in P,\, q\in Q} \Theta(p,q) < +\infty$. We assume also that $b > -\infty$; in many applications, the function $\Theta(p,q)$ is nonnegative, so this additional assumption is unnecessary. We do not always assume there are $\hat{p} \in P$ and $\hat{q} \in Q$ such that $\Theta(\hat{p},\hat{q}) = b$; when we do assume that such a $\hat{p}$ and $\hat{q}$ exist, we will not assume that $\hat{p}$ and $\hat{q}$ are unique with that property. The objective is to generate a sequence $\{(p^n,q^n)\}$ such that $\Theta(p^n,q^n) \to b$.

12.1.2 The AM Iteration

The general AM method proceeds in two steps: we begin with some q^0, and, having found q^n, we

1. minimize $\Theta(p,q^n)$ over $p \in P$ to get $p = p^{n+1}$, and then
2. minimize $\Theta(p^{n+1},q)$ over $q \in Q$ to get $q = q^{n+1}$.

In certain applications we consider the special case of alternating cross-entropy minimization. In that case, the vectors p and q are nonnegative, and the function $\Theta(p,q)$ will have the value $+\infty$ whenever there is an index j such that $p_j > 0$, but $q_j = 0$. It is important for those particular applications that we select q^0 with all positive entries. We therefore assume, for the general case, that we have selected q^0 so that $\Theta(p,q^0)$ is finite for all p.

The sequence $\{\Theta(p^n,q^n)\}$ is decreasing and bounded below by b, since we have

$$\Theta(p^n,q^n) \geq \Theta(p^{n+1},q^n) \geq \Theta(p^{n+1},q^{n+1}). \tag{12.1}$$

Therefore, the sequence $\{\Theta(p^n,q^n)\}$ converges to some $B \geq b$. Without additional assumptions, we can say little more.

We know two things:

$$\Theta(p^{n+1},q^n) - \Theta(p^{n+1},q^{n+1}) \geq 0, \tag{12.2}$$

and

$$\Theta(p^n,q^n) - \Theta(p^{n+1},q^n) \geq 0. \tag{12.3}$$

Equation 12.3 can be strengthened to

$$\Theta(p,q^n) - \Theta(p^{n+1},q^n) \geq 0. \tag{12.4}$$

We need to make these inequalities more precise.

12.1.3 The Five-Point Property for AM

The five-point property is the following: for all $p \in P$ and $q \in Q$ and $n = 1, 2, \ldots$

The Five-Point Property

$$\Theta(p, q) + \Theta(p, q^{n-1}) \geq \Theta(p, q^n) + \Theta(p^n, q^{n-1}). \tag{12.5}$$

12.1.4 The Main Theorem for AM

We want to find sufficient conditions for the sequence $\{\Theta(p^n, q^n)\}$ to converge to b, that is, for $B = b$. The following is the main result of [105].

Theorem 12.1 *If the five-point property holds then $B = b$.*

Proof: Suppose that $B > b$. Then there are p' and q' such that $B > \Theta(p', q') \geq b$. From the five-point property we have

$$\Theta(p', q^{n-1}) - \Theta(p^n, q^{n-1}) \geq \Theta(p', q^n) - \Theta(p', q'), \tag{12.6}$$

so that

$$\Theta(p', q^{n-1}) - \Theta(p', q^n) \geq \Theta(p^n, q^{n-1}) - \Theta(p', q') \geq 0. \tag{12.7}$$

All the terms being subtracted can be shown to be finite. It follows that the sequence $\{\Theta(p', q^{n-1})\}$ is decreasing, bounded below, and therefore convergent. The right side of Equation (12.7) must therefore converge to zero, which is a contradiction. We conclude that $B = b$ whenever the five-point property holds in AM. ∎

12.1.5 The Three- and Four-Point Properties

In [105] the five-point property is related to two other properties, the three- and four-point properties. This is a bit peculiar for two reasons: first, as we have just seen, the five-point property is sufficient to prove the main theorem; and second, these other properties involve a second function, $\Delta : P \times P \to [0, +\infty]$, with $\Delta(p, p) = 0$ for all $p \in P$. The three- and four-point properties jointly imply the five-point property, but to get the converse, we need to use the five-point property to define this second function; it can be done, however.

The three-point property is the following:

The Three-Point Property

$$\Theta(p, q^n) - \Theta(p^{n+1}, q^n) \geq \Delta(p, p^{n+1}), \tag{12.8}$$

for all p.

The four-point property is the following:

The Four-Point Property

$$\Delta(p, p^{n+1}) + \Theta(p, q) \geq \Theta(p, q^{n+1}), \tag{12.9}$$

for all p and q.

It is clear that the three- and four-point properties together imply the five-point property. We show now that the three-point property and the four-point property are implied by the five-point property. For that purpose we need to define a suitable $\Delta(p, \tilde{p})$. For any p and $\tilde{p}$ in P define

$$\Delta(p, \tilde{p}) = \Theta(p, q(\tilde{p})) - \Theta(p, q(p)), \tag{12.10}$$

where $q(p)$ denotes a member of Q satisfying $\Theta(p, q(p)) \leq \Theta(p, q)$, for all q in Q. Clearly, $\Delta(p, \tilde{p}) \geq 0$ and $\Delta(p, p) = 0$. The four-point property holds automatically from this definition, while the three-point property follows from the five-point property. Therefore, it is sufficient to discuss only the five-point property when speaking of the AM method.

12.1.6 Alternating Bregman Distance Minimization

The general problem of minimizing $\Theta(p, q)$ is simply a minimization of a real-valued function of two variables, $p \in P$ and $q \in Q$. In many cases the function $\Theta(p, q)$ is a distance between p and q, either $\|p-q\|_2^2$ or $KL(p, q)$. In the case of $\Theta(p, q) = \|p-q\|_2^2$, each step of the alternating minimization algorithm involves an orthogonal projection onto a closed convex set; both projections are with respect to the same Euclidean distance function. In the case of cross-entropy minimization, we first project q^n onto the set P by minimizing the distance $KL(p, q^n)$ over all $p \in P$, and then project p^{n+1} onto the set Q by minimizing the distance function $KL(p^{n+1}, q)$. This suggests the possibility of using alternating minimization with respect to more general distance functions. We shall focus on Bregman distances.

12.1.7 Bregman Distances

Let $f : \mathbb{R}^J \rightarrow \mathbb{R}$ be a Bregman function [31, 95, 39], and so $f(x)$ is convex on its domain and differentiable in the interior of its domain. Then, for x in the domain and z in the interior, we define the Bregman distance $D_f(x, z)$ by

$$D_f(x, z) = f(x) - f(z) - \langle \nabla f(z), x - z \rangle. \tag{12.11}$$

For example, the KL distance is a Bregman distance with associated Bregman function

$$f(x) = \sum_{j=1}^{J} x_j \log x_j \; - x_j. \tag{12.12}$$

Suppose now that $f(x)$ is a Bregman function and P and Q are closed convex subsets of the interior of the domain of $f(x)$. Let p^{n+1} minimize $D_f(p, q^n)$ over all $p \in P$. It follows then that

$$\langle \nabla f(p^{n+1}) - \nabla f(q^n), p - p^{n+1} \rangle \geq 0, \tag{12.13}$$

for all $p \in P$. Since

$$D_f(p, q^n) - D_f(p^{n+1}, q^n) =$$

$$D_f(p, p^{n+1}) + \langle \nabla f(p^{n+1}) - \nabla f(q^n), p - p^{n+1} \rangle, \tag{12.14}$$

it follows that the three-point property holds, with

$$\Theta(p, q) = D_f(p, q), \tag{12.15}$$

and

$$\Delta(p, \tilde{p}) = D_f(p, \tilde{p}). \tag{12.16}$$

To get the four-point property we need to restrict D_f somewhat; we assume from now on that $D_f(p, q)$ is jointly convex, that is, it is convex in the combined vector variable (p, q) (see [14]). Now we can invoke a lemma due to Eggermont and LaRiccia [117].

12.1.8 The Eggermont-LaRiccia Lemma

Lemma 12.1 *Suppose that the Bregman distance $D_f(p, q)$ is jointly convex. Then it has the four-point property.*

Proof: By joint convexity we have

$$D_f(p, q) - D_f(p^n, q^n) \geq$$

$$\langle \nabla_1 D_f(p^n, q^n), p - p^n \rangle + \langle \nabla_2 D_f(p^n, q^n), q - q^n \rangle,$$

where ∇_1 denotes the gradient with respect to the first vector variable. Since q^n minimizes $D_f(p^n, q)$ over all $q \in Q$, we have

$$\langle \nabla_2 D_f(p^n, q^n), q - q^n \rangle \geq 0,$$

for all q. Also,

$$\langle \nabla_1 D_f(p^n, q^n), p - p^n \rangle = \langle \nabla f(p^n) - \nabla f(q^n), p - p^n \rangle.$$

It follows that

$$D_f(p, q^n) - D_f(p, p^n) = D_f(p^n, q^n) + \langle \nabla_1 D_f(p^n, q^n), p - p^n \rangle$$

$$\leq D_f(p,q) - \langle \nabla_2 D_f(p^n,q^n), q-q^n \rangle \leq D_f(p,q).$$

Therefore, we have

$$D_f(p,p^n) + D_f(p,q) \geq D_f(p,q^n).$$

This is the four-point property. ∎

We now know that the alternating minimization method works for any Bregman distance that is jointly convex. This includes the Euclidean and the KL distances.

12.1.9 Minimizing a Proximity Function

We present now an example of alternating Bregman distance minimization taken from [57]. The problem is the *convex feasibility problem* (CFP), to find a member of the intersection $C \subseteq \mathbb{R}^J$ of finitely many closed convex sets C_i, $i = 1, ..., I$, or, failing that, to minimize the proximity function

$$F(x) = \sum_{i=1}^{I} D_i(\overleftarrow{P}_i x, x), \tag{12.17}$$

where f_i are Bregman functions for which D_i, the associated Bregman distance, is jointly convex, and $\overleftarrow{P}_i x$ are the *left* Bregman projection of x onto the set C_i, that is, $\overleftarrow{P}_i x \in C_i$ and $D_i(\overleftarrow{P}_i x, x) \leq D_i(z,x)$, for all $z \in C_i$. Because each D_i is jointly convex, the function $F(x)$ is convex.

The problem can be formulated as an alternating minimization, where $P \subseteq \mathbb{R}^{IJ}$ is the product set $P = C_1 \times C_2 \times ... \times C_I$. A typical member of P has the form $p = (c^1, c^2, ..., c^I)$, where $c^i \in C_i$, and $Q \subseteq \mathbb{R}^{IJ}$ is the *diagonal* subset, meaning that the elements of Q are the I-fold product of a single x; that is $Q = \{d(x) = (x, x, ..., x) \in \mathbb{R}^{IJ}\}$. We then take

$$\Theta(p,q) = \sum_{i=1}^{I} D_i(c^i, x), \tag{12.18}$$

and $\Delta(p,\tilde{p}) = \Theta(p,\tilde{p})$.

In [83] a similar iterative algorithm was developed for solving the CFP, using the same sets P and Q, but using alternating projection, rather than alternating minimization. Now it is not necessary that the Bregman distances be jointly convex. Each iteration of their algorithm involves two steps:

1. minimize $\sum_{i=1}^{I} D_i(c^i, x^n)$ over $c^i \in C_i$, obtaining $c^i = \overleftarrow{P}_i x^n$, and then

2. minimize $\sum_{i=1}^{I} D_i(x, \overleftarrow{P}_i x^n)$.

Because this method is an alternating projection approach, it converges only when the CFP has a solution, whereas the previous alternating minimization method minimizes $F(x)$, even when the CFP has no solution.

12.1.10 Right and Left Projections

Because Bregman distances D_f are not generally symmetric, we can speak of *right* and *left* Bregman projections onto a closed convex set. For any allowable vector x, the *left* Bregman projection of x onto C, if it exists, is the vector $\overleftarrow{P}_C x \in C$ satisfying the inequality $D_f(\overleftarrow{P}_C x, x) \leq D_f(c, x)$, for all $c \in C$. Similarly, the *right* Bregman projection is the vector $\overrightarrow{P}_C x \in C$ satisfying the inequality $D_f(x, \overrightarrow{P}_C x) \leq D_f(x, c)$, for any $c \in C$.

The alternating minimization approach described above to minimize the proximity function

$$F(x) = \sum_{i=1}^{I} D_i(\overleftarrow{P}_i x, x) \tag{12.19}$$

can be viewed as an alternating projection method, but employing both right and left Bregman projections.

Consider the problem of finding a member of the intersection of two closed convex sets C and D. We could proceed as follows: having found x^n, minimize $D_f(x^n, d)$ over all $d \in D$, obtaining $d = \overrightarrow{P}_D x^n$, and then minimize $D_f(c, \overrightarrow{P}_D x^n)$ over all $c \in C$, obtaining $c = x^{n+1} = \overleftarrow{P}_C \overrightarrow{P}_D x^n$. The objective of this algorithm is to minimize $D_f(c, d)$ over all $c \in C$ and $d \in D$; such a minimum may not exist, of course.

In [16] the authors note that the alternating minimization algorithm of [57] involves right and left Bregman projections, which suggests to them iterative methods involving a wider class of operators that they call "Bregman retractions".

12.1.11 More Proximity Function Minimization

Proximity function minimization and right and left Bregman projections play a role in a variety of iterative algorithms. We survey several of them in this section.

12.1.12 Cimmino's Algorithm

Our objective here is to find an exact or approximate solution of the system of I linear equations in J unknowns, written $Ax = b$. For each i let

$$C_i = \{z | (Az)_i = b_i\}, \tag{12.20}$$

and $P_i x$ be the orthogonal projection of x onto C_i. Then

$$(P_i x)_j = x_j + \alpha_i A_{ij}(b_i - (Ax)_i), \tag{12.21}$$

where

$$(\alpha_i)^{-1} = \sum_{j=1}^{J} A_{ij}^2. \tag{12.22}$$

Let

$$F(x) = \sum_{i=1}^{I} \|P_i x - x\|_2^2. \tag{12.23}$$

Using alternating minimization on this proximity function gives Cimmino's algorithm, with the iterative step

$$x_j^k = x_j^{k-1} + \frac{1}{I}\sum_{i=1}^{I} \alpha_i A_{ij}(b_i - (Ax^{k-1})_i). \tag{12.24}$$

12.1.13 Simultaneous Projection for Convex Feasibility

Now we let C_i be any closed convex subsets of $\mathbb{R}^J$ and define $F(x)$ as in the previous section. Again, we apply alternating minimization. The iterative step of the resulting algorithm is

$$x^k = \frac{1}{I}\sum_{i=1}^{I} P_i x^{k-1}. \tag{12.25}$$

The objective here is to minimize $F(x)$, if there is a minimum.

12.1.14 The Bauschke-Combettes-Noll Problem

In [17] Bauschke, Combettes and Noll consider the following problem: minimize the function

$$\Theta(p,q) = \Lambda(p,q) = \phi(p) + \psi(q) + D_f(p,q), \tag{12.26}$$

where ϕ and ψ are convex on $\mathbb{R}^J$, $D = D_f$ is a Bregman distance, and $P = Q$ is the interior of the domain of f. They assume that

$$b = \inf_{(p,q)} \Lambda(p,q) > -\infty, \tag{12.27}$$

and seek a sequence $\{(p^n, q^n)\}$ such that $\{\Lambda(p^n, q^n)\}$ converges to b. The sequence is obtained by the AM method, as in our previous discussion. They prove that, if the Bregman distance is jointly convex, then $\{\Lambda(p^n, q^n)\} \downarrow b$. In this subsection we obtain this result by showing that $\Lambda(p, q)$ has the five-point property whenever $D = D_f$ is jointly convex. Our proof is loosely based on the proof of the Eggermont-LaRiccia lemma.

The five-point property for $\Lambda(p, q)$ is

$$\Lambda(p, q^{n-1}) - \Lambda(p^n, q^{n-1}) \geq \Lambda(p, q^n) - \Lambda(p, q). \tag{12.28}$$

Lemma 12.2 *The inequality in (12.28) is equivalent to*

$$\Lambda(p, q) - \Lambda(p^n, q^n) \geq$$
$$D(p, q^n) + D(p^n, q^{n-1}) - D(p, q^{n-1}) - D(p^n, q^n). \tag{12.29}$$

Proof: The proof is Exercise 12.1. ∎

By the joint convexity of $D(p, q)$ and the convexity of ϕ and ψ we have

$$\Lambda(p, q) - \Lambda(p^n, q^n) \geq$$
$$\langle \nabla_p \Lambda(p^n, q^n), p - p^n \rangle + \langle \nabla_q \Lambda(p^n, q^n), q - q^n \rangle, \tag{12.30}$$

where $\nabla_p \Lambda(p^n, q^n)$ denotes the gradient of $\Lambda(p, q)$, with respect to p, evaluated at (p^n, q^n).

Since q^n minimizes $\Lambda(p^n, q)$, it follows that

$$\langle \nabla_q \Lambda(p^n, q^n), q - q^n \rangle = 0, \tag{12.31}$$

for all q. Therefore,

$$\Lambda(p, q) - \Lambda(p^n, q^n) \geq \langle \nabla_p \Lambda(p^n, q^n), p - p^n \rangle. \tag{12.32}$$

We have

$$\langle \nabla_p \Lambda(p^n, q^n), p - p^n \rangle =$$
$$\langle \nabla f(p^n) - \nabla f(q^n), p - p^n \rangle + \langle \nabla \phi(p^n), p - p^n \rangle. \tag{12.33}$$

Since p^n minimizes $\Lambda(p, q^{n-1})$, we have

$$\nabla_p \Lambda(p^n, q^{n-1}) = 0, \tag{12.34}$$

or

$$\nabla \phi(p^n) = \nabla f(q^{n-1}) - \nabla f(p^n), \tag{12.35}$$

so that

$$\langle \nabla_p \Lambda(p^n, q^n), p - p^n \rangle = \langle \nabla f(q^{n-1}) - \nabla f(q^n), p - p^n \rangle$$
$$= D(p, q^n) + D(p^n, q^{n-1}) - D(p, q^{n-1}) - D(p^n, q^n). \tag{12.36}$$

Using (12.32) we obtain the inequality in (12.29). This shows that $\Lambda(p, q)$ has the five-point property whenever the Bregman distance $D = D_f$ is jointly convex. From our previous discussion of AM, we conclude that the sequence $\{\Lambda(p^n, q^n)\}$ converges to b; this is Corollary 4.3 of [17].

As we shall see in the next chapter, the expectation maximization maximum likelihood (EM) method involves alternating minimization of a function of the form $\Lambda(p, q)$.

If $\psi = 0$, then $\{\Lambda(p^n, q^n)\}$ converges to b, even without the assumption that the distance D_f is jointly convex. In such cases, $\Lambda(p, q)$ has the form of the objective function in proximal minimization and therefore the problem falls into the SUMMA class (see Lemma 4.1).

12.1.15 AM as SUMMA

We show now that the SUMMA class of sequential unconstrained minimization methods includes all the AM methods for which the five-point property holds.

For each p in the set P, define $q(p)$ in Q as a member of Q for which $\Theta(p, q(p)) \leq \Theta(p, q)$, for all $q \in Q$. Let $f(p) = \Theta(p, q(p))$.

At the nth step of AM we minimize

$$G_n(p) = \Theta(p, q^{n-1}) = \Theta(p, q(p)) + \Big(\Theta(p, q^{n-1}) - \Theta(p, q(p))\Big) \quad (12.37)$$

to get p^n. With

$$g_n(p) = \Big(\Theta(p, q^{n-1}) - \Theta(p, q(p))\Big) \geq 0, \quad (12.38)$$

we can write

$$G_n(p) = f(p) + g_n(p). \quad (12.39)$$

According to the five-point property, we have

$$G_n(p) - G_n(p^n) \geq \Theta(p, q^n) - \Theta(p, q(p)) = g_{n+1}(p). \quad (12.40)$$

It follows that AM is a member of the SUMMA class.

12.2 Exercises

Ex. 12.1 *Prove Lemma 12.2.*

Chapter 13

The EM Algorithm

13.1 Overview

The "expectation maximization" (EM) algorithm is a general framework for maximizing the likelihood function in statistical parameter estimation [167]. The EM algorithm is not really a single algorithm, but a framework for the design of iterative likelihood maximization methods, or, as the authors of [21] put it, a "prescription for constructing an algorithm"; nevertheless, we shall continue to refer to *the* EM algorithm. We show in this chapter that EM algorithms are AF algorithms. The EM algorithms are always presented in probabilistic terms, involving the maximization of a conditional expected value. As we shall demonstrate, the essence of the EM algorithm is not stochastic. Our non-stochastic EM (NSEM) is a general approach for function maximization that has the stochastic EM methods as particular cases.

Maximizing the likelihood function is a well studied procedure for estimating parameters from observed data. When a maximizer cannot be obtained in closed form, iterative maximization algorithms, such as the expectation maximization (EM) maximum likelihood algorithms, are needed. The standard formulation of the EM algorithms postulates that finding a maximizer of the likelihood is complicated because the observed data is somehow incomplete or deficient, and the maximization would have been simpler had we observed the complete data. The EM algorithm involves repeated calculations involving complete data that has been estimated using the current parameter value and conditional expectation.

The standard formulation is adequate for the most common discrete case, in which the random variables involved are governed by finite or infinite probability functions, but unsatisfactory in general, particularly in the continuous case, in which probability density functions and integrals are needed.

We adopt the view that the observed data is not necessarily incomplete, but just difficult to work with, while different data, which we call the preferred data, leads to simpler calculations. To relate the preferred data to the observed data, we assume that the preferred data is *acceptable*, which means that the conditional distribution of the preferred data, given the observed data, is independent of the parameter. This extension of the EM algorithms contains the usual formulation for the discrete case, while removing the difficulties associated with the continuous case. Examples are given to illustrate this new approach.

13.2 A Non-Stochastic Formulation of EM

The essence of the EM algorithm is not stochastic, and leads to a general approach for function maximization, which we call the "non-stochastic" EM algorithm (NSEM)[67]. In addition to being more general, this new approach also simplifies much of the development of the EM algorithm itself. We present now the essential aspects of the EM algorithm without relying on statistical concepts. We shall use these results later to establish important facts about the statistical EM algorithm.

13.2.1 The Continuous Case

The problem is to maximize a nonnegative function $f : Z \to \mathbb{R}$, where Z is an arbitrary set. We assume that there is $z^* \in Z$ with $f(z^*) \geq f(z)$, for all $z \in Z$. We also assume that there is a nonnegative function $b : \mathbb{R}^J \times Z \to \mathbb{R}$ such that

$$f(z) = \int b(x, z)dx.$$

Having found z^k, we maximize the function

$$H(z^k, z) = \int b(x, z^k) \log b(x, z)dx \tag{13.1}$$

to get z^{k+1}. Adopting such an iterative approach presupposes that maximizing $H(z^k, z)$ is simpler than maximizing $f(z)$ itself. This is the case with the EM algorithm.

The cross-entropy or Kullback-Leibler distance [153] is a useful tool for analyzing the EM algorithm. We simplify the notation by setting $b(z) = b(x, z)$. Maximizing $H(z^k, z)$ is equivalent to minimizing

$$G(z^k, z) = KL(b(z^k), b(z)) - f(z), \tag{13.2}$$

where

$$KL(b(z^k), b(z)) = \int KL(b(x, z^k), b(x, z))dx. \tag{13.3}$$

Therefore,

$$-f(z^k) = KL(b(z^k), b(z^k)) - f(z^k) \geq KL(b(z^k), b(z^{k+1})) - f(z^{k+1}),$$

or

$$f(z^{k+1}) - f(z^k) \geq KL(b(z^k), b(z^{k+1})) \geq KL(f(z^k), f(z^{k+1})).$$

Consequently, the sequence $\{f(z^k)\}$ is increasing and bounded above, so that the sequence $\{KL(b(z^k), b(z^{k+1}))\}$ converges to zero. Without additional restrictions, we cannot conclude that $\{f(z^k)\}$ converges to $f(z^*)$.

We get z^{k+1} by minimizing $G(z^k, z)$. When we minimize $G(z, z^{k+1})$, we get z^{k+1} again. Therefore, we can put the NSEM algorithm into the alternating minimization (AM) framework of Csiszár and Tusnády [105], as discussed in Chapter 12.

13.2.2 The Discrete Case

Again, the problem is to maximize a nonnegative function $f : Z \to \mathbb{R}$, where Z is an arbitrary set. As previously, we assume that there is $z^* \in Z$

with $f(z^*) \geq f(z)$, for all $z \in Z$. We also assume that there is a finite or countably infinite set B and a nonnegative function $b : B \times Z \to \mathbb{R}$ such that

$$f(z) = \sum_{x \in B} b(x, z).$$

Having found z^k, we maximize the function

$$H(z^k, z) = \sum_{x \in B} b(x, z^k) \log b(x, z) \tag{13.4}$$

to get z^{k+1}.

We set $b(z) = b(x, z)$ again. Maximizing $H(z^k, z)$ is equivalent to minimizing

$$G(z^k, z) = KL(b(z^k), b(z)) - f(z), \tag{13.5}$$

where

$$KL(b(z^k), b(z)) = \sum_{x \in B} KL(b(x, z^k), b(x, z)). \tag{13.6}$$

As previously, we find that $\{f(z^k)\}$ is increasing, and $\{KL(b(z^k), b(z^{k+1}))\}$ converges to zero. Without additional restrictions, we cannot conclude that $\{f(z^k)\}$ converges to $f(z^*)$.

13.3 The Stochastic EM Algorithm

In this section we present the standard stochastic formulation of the EM algorithm.

13.3.1 The E-step and M-step

In statistical parameter estimation one typically has an *observable* random vector Y taking values in $\mathbb{R}^N$ that is governed by a probability density function (pdf) or probability function (pf) of the form $f_Y(y|\theta)$, for some value of the parameter vector $\theta \in \Theta$, where Θ is the set of all legitimate values of θ. Our *observed* data consists of one realization y of Y; we do not exclude the possibility that the entries of y are independently obtained samples of a common real-valued random variable. The true vector of parameters is to be estimated by maximizing the likelihood function $L_y(\theta) = f_Y(y|\theta)$ over all $\theta \in \Theta$ to obtain a maximum likelihood estimate, θ_{ML}.

To employ the EM algorithmic approach, it is assumed that there is another related random vector X, which we shall call the *preferred* data, such that, had we been able to obtain one realization x of X, maximizing the likelihood function $L_x(\theta) = f_X(x|\theta)$ would have been simpler than maximizing the likelihood function $L_y(\theta) = f_Y(y|\theta)$. Of course, we do not have a realization x of X. The basic idea of the EM approach is to estimate x using conditional expectations and the current estimate of θ, denoted θ^k, and to use each estimate x^k of x to get the next estimate θ^{k+1}.

The EM algorithm proceeds in two steps. Having selected the preferred data X, and having found θ^k, we form the function of θ given by the conditional expected value

$$Q(\theta|\theta^k) = E(\log f_X(x|\theta)|y, \theta^k); \tag{13.7}$$

this is the E-step of the EM algorithm. Then we maximize $Q(\theta|\theta^k)$ over all θ to get θ^{k+1}; this is the M-step of the EM algorithm. In this way, the EM algorithm based on X generates a sequence $\{\theta^k\}$ of parameter vectors.

For the discrete case of probability functions, we have

$$Q(\theta|\theta^k) = \sum_x f_{X|Y}(x|y, \theta^k) \log f_X(x|\theta), \tag{13.8}$$

and for the continuous case of probability density functions we have

$$Q(\theta|\theta^k) = \int f_{X|Y}(x|y, \theta^k) \log f_X(x|\theta) dx. \tag{13.9}$$

In decreasing order of importance and difficulty, the goals are these:

1. to have the sequence of parameters $\{\theta^k\}$ converging to θ_{ML};
2. to have the sequence of functions $\{f_X(x|\theta^k)\}$ converging to $f_X(x|\theta_{ML})$;
3. to have the sequence of numbers $\{L_y(\theta^k)\}$ converging to $L_y(\theta_{ML})$;
4. to have the sequence of numbers $\{L_y(\theta^k)\}$ non-decreasing.

Our focus here is mainly on the fourth goal, with some discussion of the third goal. We do present some examples for which all four goals are attained. Clearly, the first goal requires a topology on the set Θ.

13.3.2 Difficulties with the Conventional Formulation

In [167] we are told that

$$f_{X|Y}(x|y, \theta) = f_X(x|\theta)/f_Y(y|\theta). \tag{13.10}$$

This is false; integrating with respect to x gives 1 on the left side and $1/f_Y(y|\theta)$ on the right side. Perhaps the equation is not meant to hold for all x, but just for some x. In fact, if there is a function h such that $Y = h(X)$, then Equation (13.10) might hold for those x such that $h(x) = y$. In fact, this is what happens in the discrete case of probabilities; in that case we do have

$$f_Y(y|\theta) = \sum_{x \in h^{-1}\{y\}} f_X(x|\theta), \tag{13.11}$$

where

$$h^{-1}\{y\} = \{x|h(x) = y\}.$$

Consequently,

$$f_{X|Y}(x|y,\theta) = f_X(x|\theta)/f_Y(y|\theta), \text{ if } x \in h^{-1}\{y\}, \tag{13.12}$$

and zero, otherwise. However, this modification of Equation (13.10) fails in the continuous case of probability density functions, since $h^{-1}\{y\}$ is often a subset of zero measure. Even if the set $h^{-1}\{y\}$ has positive measure, integrating both sides of Equation (13.10) over $x \in h^{-1}\{y\}$ tells us that $f_Y(y|\theta) \leq 1$, which need not hold for probability density functions.

13.3.3 An Incorrect Proof

Everyone who works with the EM algorithm will say that the likelihood is non-decreasing for the EM algorithm. The proof of this fact usually proceeds as follows; we use the notation for the continuous case, but the proof for the discrete case is essentially the same. Use Equation (13.10) to get

$$\log f_X(x|\theta) = \log f_{X|Y}(x|y,\theta) - \log f_Y(y|\theta). \tag{13.13}$$

Then replace the term $\log f_X(x|\theta)$ in Equation (13.9) with the right side of Equation (13.13), obtaining

$$\log f_Y(y|\theta) - Q(\theta|\theta^k) = -\int f_{X|Y}(x|y,\theta^k) \log f_{X|Y}(x|y,\theta)dx. \tag{13.14}$$

Jensen's Inequality tells us that

$$\int u(x) \log u(x)dx \geq \int u(x) \log v(x)dx, \tag{13.15}$$

for any probability density functions $u(x)$ and $v(x)$. Since $f_{X|Y}(x|y,\theta)$ is a probability density function, we have

$$\int f_{X|Y}(x|y,\theta^k) \log f_{X|Y}(x|y,\theta)dx \leq \int f_{X|Y}(x|y,\theta^k) \log f_{X|Y}(x|y,\theta^k)dx.$$

We conclude, therefore, that $\log f_Y(y|\theta) - Q(\theta|\theta^k)$ attains its minimum value at $\theta = \theta^k$. Then we have

$$\log f_Y(y|\theta^{k+1}) - \log f_Y(y|\theta^k) \geq Q(\theta^{k+1}|\theta^k) - Q(\theta^k|\theta^k) \geq 0. \quad (13.16)$$

This proof is incorrect; clearly it rests on the validity of Equation (13.10), which is generally false. For the discrete case, with $Y = h(X)$, this proof is valid, when we use Equation (13.12), instead of Equation (13.10). In all other cases, however, the proof is incorrect.

13.3.4 Acceptable Data

We turn now to the question of how to repair the incorrect proof. Equation (13.10) should read

$$f_{X|Y}(x|y,\theta) = f_{X,Y}(x,y|\theta)/f_Y(y|\theta), \quad (13.17)$$

for all x. In order to replace $\log f_X(x|\theta)$ in Equation (13.9) we write

$$f_{X,Y}(x,y|\theta) = f_{X|Y}(x|y,\theta)f_Y(y|\theta), \quad (13.18)$$

and

$$f_{X,Y}(x,y|\theta) = f_{Y|X}(y|x,\theta)f_X(x|\theta), \quad (13.19)$$

so that

$$\log f_X(x|\theta) = \log f_{X|Y}(x|y,\theta) + \log f_Y(y|\theta) - \log f_{Y|X}(y|x,\theta). \quad (13.20)$$

We say that the preferred data is *acceptable* if

$$f_{Y|X}(y|x,\theta) = f_{Y|X}(y|x); \quad (13.21)$$

that is, the dependence of Y on X is unrelated to the value of the parameter θ. This definition provides our generalization of the relationship $Y = h(X)$.

When X is acceptable, we have that $\log f_Y(y|\theta) - Q(\theta|\theta^k)$ again attains its minimum value at $\theta = \theta^k$. The assertion that the likelihood is non-decreasing then follows, using the same argument as in the previous incorrect proof.

13.4 The Discrete Case

In the discrete case, we assume that Y is a discrete random vector taking values in a finite or countably infinite set A, and governed by probability

$f_Y(y|\theta)$. We assume, in addition, that there is a second discrete random vector X, taking values in a finite or countably infinite set B, and a function $h : B \to A$ such that $Y = h(X)$. We define the set

$$h^{-1}\{y\} = \{x \in B | h(x) = y\}. \tag{13.22}$$

Then we have

$$f_Y(y|\theta) = \sum_{x \in h^{-1}\{y\}} f_X(x|\theta). \tag{13.23}$$

The conditional probability function for X, given $Y = y$, is

$$f_{X|Y}(x|y,\theta) = \frac{f_X(x|\theta)}{f_Y(y|\theta)}, \tag{13.24}$$

for $x \in h^{-1}\{y\}$, and zero, otherwise. The so-called E-step of the EM algorithm is then to calculate $Q(\theta|\theta^k) = E((\log f_X(X|\theta)|y, \theta^k)$ where

$$E((\log f_X(X|\theta)|y,\theta^k) = \sum_{x \in h^{-1}\{y\}} f_{X|Y}(x|y,\theta^k) \log f_X(x|\theta), \tag{13.25}$$

and the M-step is to maximize $Q(\theta|\theta^k)$ as a function of θ to obtain θ^{k+1}.

Using Equation (13.24), we can write

$$Q(\theta|\theta^k) = \sum_{x \in h^{-1}\{y\}} f_{X|Y}(x|y,\theta^k) \log f_{X|Y}(x|y,\theta) + \log f_Y(y|\theta). \tag{13.26}$$

Therefore,

$$\log f_Y(y|\theta) - Q(\theta|\theta^k) = - \sum_{x \in h^{-1}\{y\}} f_{X|Y}(x|y,\theta^k) \log f_{X|Y}(x|y,\theta).$$

Since

$$\sum_{x \in h^{-1}\{y\}} f_{X|Y}(x|y,\theta^k) = \sum_{x \in h^{-1}\{y\}} f_{X|Y}(x|y,\theta) = 1,$$

that

$$- \sum_{x \in h^{-1}\{y\}} f_{X|Y}(x|y,\theta^k) \log f_{X|Y}(x|y,\theta) \geq$$

$$- \sum_{x \in h^{-1}\{y\}} f_{X|Y}(x|y,\theta^k) \log f_{X|Y}(x|y,\theta^k)$$

follows from Jensen's Inequality. Therefore, $\log f_Y(y|\theta) - Q(\theta|\theta^k)$ attains its minimum at $\theta = \theta^k$. We have the following result.

Proposition 13.1 *The sequence $\{f_Y(y|\theta^k)\}$ is non-decreasing.*

Proof: We have

$$\log f_Y(y|\theta^{k+1}) - Q(\theta^{k+1}|\theta^k) \geq \log f_Y(y|\theta^k) - Q(\theta^k|\theta^k),$$

or

$$\log f_Y(y|\theta^{k+1}) - \log f_Y(y|\theta^k) \geq Q(\theta^{k+1}|\theta^k) - Q(\theta^k|\theta^k) \geq 0.$$

∎

Let $\chi_{h^{-1}\{y\}}(x)$ be the characteristic function of the set $h^{-1}\{y\}$, that is,

$$\chi_{h^{-1}\{y\}}(x) = 1,$$

for $x \in h^{-1}\{y\}$, and zero, otherwise. With the choices $z = \theta$, $f(z) = f_Y(y|\theta)$, and $b(z) = f_X(x|\theta)\chi_{h^{-1}\{y\}}(x)$, the discrete EM algorithm fits into the framework of the non-stochastic EM algorithm. Consequently, we see once again that the sequence $\{f_Y(y|\theta^k)\}$ is non-decreasing, and also that the sequence

$$\{KL(b(z^k), b(z^{k+1}))\} = \{\sum_{x \in h^{-1}\{y\}} KL(f_X(x|\theta^k), f_X(x|\theta^{k+1}))\}$$

converges to zero.

13.5 Missing Data

We say that there is *missing data* if the preferred data X has the form $X = (Y, W)$, so that $Y = h(X) = h(Y, W)$, where h is the orthogonal projection onto the first component. The case of missing data for the discrete case is covered by the discussion in Section 13.4, so we consider here the continuous case in which probability density functions are involved.

Once again, the E-step is to calculate $Q(\theta|\theta^k)$ given by

$$Q(\theta|\theta^k) = E((\log f_X(X|\theta)|y, \theta^k). \tag{13.27}$$

Since $X = (Y, W)$, we have

$$f_X(x|\theta) = f_{Y,W}(y, w|\theta). \tag{13.28}$$

Since the set $h^{-1}\{y\}$ has measure zero, we cannot write

$$Q(\theta|\theta^k) = \int_{h^{-1}\{y\}} f_{X|Y}(x|y, \theta^k) \log f_X(x|\theta) dx.$$

Instead, we write

$$Q(\theta|\theta^k) = \int f_{Y,W}(y,w|\theta^k)\log f_{Y,W}(y,w|\theta)dw/f_Y(y|\theta^k). \tag{13.29}$$

Consequently, maximizing $Q(\theta|\theta^k)$ is equivalent to maximizing

$$\int f_{Y,W}(y,w|\theta^k)\log f_{Y,W}(y,w|\theta)dw.$$

With $b(\theta) = b(\theta, w) = f_{Y,W}(y, w|\theta)$ and

$$f_Y(y|\theta) = f(\theta) = \int f_{Y,W}(y,w|\theta)dw = \int b(\theta)dw,$$

we find that maximizing $Q(\theta|\theta^k)$ is equivalent to minimizing $KL(b(\theta^k), b(\theta)) - f(\theta)$. Therefore, the EM algorithm for the case of missing data falls into the framework of the non-stochastic EM algorithm. We conclude that the sequence $\{f(\theta^k)\}$ is non-decreasing, and that the sequence $\{KL(b(\theta^k), b(\theta^{k+1}))\}$ converges to zero.

Most other instances of the continuous case in which we have $Y = h(X)$ can be handled using the missing-data model. For example, suppose that Z_1 and Z_2 are uniformly distributed on the interval $[0, \theta]$, for some positive θ, and that $Y = Z_1+Z_2$. We may, for example, then take W to be $W = Z_1-Z_2$ and $X = (Y, W)$ as the preferred data.

13.6 The Continuous Case

We turn now to the general continuous case. We have a random vector Y taking values in $\mathbb{R}^J$ and governed by the probability density function $f_Y(y|\theta)$. The objective, once again, is to maximize the likelihood function $L_y(\theta) = f_Y(y|\theta)$ to obtain the maximum likelihood estimate of θ.

13.6.1 Acceptable Preferred Data

For the continuous case, the vector θ^{k+1} is obtained from θ^k by maximizing the conditional expected value

$$Q(\theta|\theta^k) = E(\log f_X(X|\theta)|y, \theta^k) = \int f_{X|Y}(x|y,\theta^k)\log f_X(x|\theta)dx. \tag{13.30}$$

Assuming the acceptability condition and using

$$f_{X,Y}(x, y|\theta^k) = f_{X|Y}(x|y, \theta^k)f_Y(y|\theta^k),$$

and

$$\log f_X(x|\theta) = \log f_{X,Y}(x,y|\theta) - \log f_{Y|X}(y|x),$$

we find that maximizing $E(\log f_X(x|\theta)|y,\theta^k)$ is equivalent to minimizing

$$H(\theta^k,\theta) = \int f_{X,Y}(x,y|\theta^k)\log f_{X,Y}(x,y|\theta)dx. \tag{13.31}$$

With $f(\theta) = f_Y(y|\theta)$, and $b(\theta) = f_{X,Y}(x,y|\theta)$, this problem fits the framework of the non-stochastic EM algorithm and is equivalent to minimizing

$$G(\theta^k,\theta) = KL(b(\theta^k), b(\theta)) - f(\theta).$$

Once again, we may conclude that the likelihood function is non-decreasing and that the sequence $\{KL(b(\theta^k), b(\theta^{k+1}))\}$ converges to zero.

In the discrete case in which $Y = h(X)$ the conditional probability $f_{Y|X}(y|x,\theta)$ is $\delta(y - h(x))$, as a function of y, for given x, and is the characteristic function of the set $\mathcal{X}(y)$, as a function of x, for given y. Therefore, we can write $f_{X|Y}(x|y,\theta)$ using Equation (13.12). For the continuous case in which $Y = h(X)$, the pdf $f_{Y|X}(y|x,\theta)$ is again a delta function of y, for given x; the difficulty arises when we need to view this as a function of x, for given y. The acceptability property helps us avoid this difficulty.

When X is acceptable, we have

$$f_{X|Y}(x|y,\theta) = f_{Y|X}(y|x)f_X(x|\theta)/f_Y(y|\theta), \tag{13.32}$$

whenever $f_Y(y|\theta) \neq 0$, and is zero otherwise. Consequently, when X is acceptable, we have a kernel model for $f_Y(y|\theta)$ in terms of the $f_X(x|\theta)$:

$$f_Y(y|\theta) = \int f_{Y|X}(y|x)f_X(x|\theta)dx; \tag{13.33}$$

for the continuous case we view this as a corrected version of Equation (13.11). In the discrete case the integral is replaced by a summation, of course, but when we are speaking generally about either case, we shall use the integral sign.

The acceptability of the missing data W is used in [68], but more for computational convenience and to involve the Kullback-Leibler distance in the formulation of the EM algorithm. It is not necessary that W be acceptable in order for likelihood to be non-decreasing, as we have seen.

13.6.2 Selecting Preferred Data

The popular example of multinomial data given below illustrates well the point that one can often choose to view the observed data as "incomplete" simply in order to introduce "complete" data that makes the calculations simpler, even when there is no suggestion, in the original problem, that the observed data is in any way inadequate or "incomplete". It is in order to emphasize this desire for simplification that we refer to X as the preferred data, not the complete data.

In some applications, the preferred data X arises naturally from the problem, while in other cases the user must imagine preferred data. This choice in selecting the preferred data can be helpful in speeding up the algorithm (see [125]).

If, instead of maximizing

$$\int f_{X|Y}(x|y,\theta^k)\log f_X(x|\theta)dx,$$

at each M-step, we simply select θ^{k+1} so that

$$\int f_{X|Y}(x|y,\theta^k)\log f_{X,Y}(x,y|\theta^{k+1})dx - \int f_{X|Y}(x|y,\theta^k)\log f_{X,Y}(x,y|\theta^k)dx > 0,$$

we say that we are using a *generalized* EM (GEM) algorithm. It is clear from the discussion in the previous subsection that, whenever X is acceptable, a GEM also guarantees that likelihood is non-decreasing.

13.6.3 Preferred Data as Missing Data

As we have seen, when the EM algorithm is applied to the missing-data model, the likelihood is non-decreasing, which suggests that, for an arbitrary preferred data X, we could imagine X as W, the missing data, and imagine applying the EM algorithm to $Z = (Y, X)$. This approach would produce an EM sequence of parameter vectors for which likelihood is non-decreasing, but it need not be the same sequence as obtained by applying the EM algorithm to X directly. It is the same sequence, provided that X is acceptable. We are not suggesting that applying the EM algorithm to $Z = (Y, X)$ would simplify calculations.

We know that, when the missing-data model is used and the M-step is defined as maximizing the function in (13.29), the likelihood is not decreasing. It would seem then that, for any choice of preferred data X, we could view this data as missing and take as our complete data the pair $Z = (Y, X)$, with X now playing the role of W. Maximizing the function in (13.29) is then equivalent to maximizing

$$\int f_{X|Y}(x|y,\theta^k)\log f_{X,Y}(x,y|\theta)dx; \tag{13.34}$$

to get θ^{k+1}. It then follows that $L_y(\theta^{k+1}) \geq L_y(\theta^k)$. The obvious question is whether or not these two functions given in (13.7) and (13.34) have the same maximizers.

For acceptable X we have

$$\log f_{X,Y}(x, y|\theta) = \log f_X(x|\theta) + \log f_{Y|X}(y|x), \tag{13.35}$$

so the two functions given in (13.7) and (13.34) do have the same maximizers. It follows once again that, whenever the preferred data is acceptable, we have $L_y(\theta^{k+1}) \geq L_y(\theta^k)$. Without additional assumptions, however, we cannot conclude that $\{\theta^k\}$ converges to θ_{ML}, nor that $\{f_Y(y|\theta^k)\}$ converges to $f_Y(y|\theta_{ML})$.

13.7 EM and the KL Distance

We illustrate the usefulness of acceptability and reformulate the M-step in terms of cross-entropy or Kullback-Leibler distance minimization.

13.7.1 Using Acceptable Data

The assumption that the data X is acceptable helps simplify the theoretical discussion of the EM algorithm.

For any preferred X the M-step of the EM algorithm, in the continuous case, is to maximize the function

$$\int f_{X|Y}(x|y, \theta^k) \log f_X(x|\theta) dx, \tag{13.36}$$

over $\theta \in \Theta$; the integral is replaced by a sum in the discrete case. For notational convenience we let

$$b(\theta^k) = f_{X|Y}(x|y, \theta^k), \tag{13.37}$$

and

$$f(\theta) = f_X(x|\theta); \tag{13.38}$$

both functions are functions of the vector variable x. Then the M-step is equivalent to minimizing the Kullback-Leibler or cross-entropy distance

$$KL(b(\theta^k), f(\theta)) = \int f_{X|Y}(x|y, \theta^k) \log \left(\frac{f_{X|Y}(x|y,\theta^k)}{f_X(x|\theta)}\right) dx$$

$$= \int f_{X|Y}(x|y, \theta^k) \log \left(\frac{f_{X|Y}(x|y,\theta^k)}{f_X(x|\theta)}\right) + f_X(x|\theta) - f_{X|Y}(x|y, \theta^k) dx. \tag{13.39}$$

This holds since both $f_X(x|\theta)$ and $f_{X|Y}(x|y,\theta^k)$ are probability density functions or probabilities.

For acceptable X we have

$$\log f_{X,Y}(x,y|\theta) = \log f_{X|Y}(x|y,\theta) + \log f_Y(y|\theta)$$
$$= \log f_{Y|X}(y|x) + \log f_X(x|\theta). \tag{13.40}$$

Therefore,

$$\log f_Y(y|\theta^{k+1}) - \log f_Y(y|\theta) =$$
$$KL(b(\theta^k), f(\theta)) - KL(b(\theta^k), f(\theta^{k+1}))$$
$$+KL(b(\theta^k), b(\theta^{k+1})) - KL(b(\theta^k), b(\theta)). \tag{13.41}$$

Since $\theta = \theta^{k+1}$ minimizes $KL(b(\theta^k), f(\theta))$, we have that

$$\log f_Y(y|\theta^{k+1}) - \log f_Y(y|\theta^k) =$$
$$KL(b(\theta^k), f(\theta^k)) - KL(b(\theta^k), f(\theta^{k+1})) + KL(b(\theta^k), b(\theta^{k+1})) \geq 0. \tag{13.42}$$

This tells us, again, that the sequence of likelihood values $\{\log f_Y(y|\theta^k)\}$ is increasing, and that the sequence of its negatives, $\{-\log f_Y(y|\theta^k)\}$, is decreasing. Since we assume that there is a maximizer θ_{ML} of the likelihood, the sequence $\{-\log f_Y(y|\theta^k)\}$ is also bounded below and the sequences $\{KL(b(\theta^k), b(\theta^{k+1}))\}$ and $\{KL(b(\theta^k), f(\theta^k)) - KL(b(\theta^k), f(\theta^{k+1}))\}$ converge to zero.

Without some notion of convergence in the parameter space Θ, we cannot conclude that $\{\theta^k\}$ converges to a maximum likelihood estimate θ_{ML}. Without some additional assumptions, we cannot even conclude that the functions $f(\theta^k)$ converge to $f(\theta_{ML})$.

13.8 Finite Mixture Problems

Estimating the combining proportions in probabilistic mixture problems shows that there are meaningful examples of our acceptable-data model, and provides important applications of likelihood maximization.

13.8.1 Mixtures

We say that a random vector V taking values in $\mathbb{R}^D$ is a *finite mixture* (see [119, 187]) if there are probability density functions or probabilities f_j and numbers $\theta_j \geq 0$, for $j = 1, ..., J$, such that the probability density function or probability function for V has the form

$$f_V(v|\theta) = \sum_{j=1}^{J} \theta_j f_j(v), \tag{13.43}$$

for some choice of the $\theta_j \geq 0$ with $\sum_{j=1}^{J} \theta_j = 1$. As previously, we shall assume, without loss of generality, that $D = 1$.

13.8.2 The Likelihood Function

The data are N realizations of the random variable V, denoted v_n, for $n = 1, ..., N$, and the given data is the vector $y = (v_1, ..., v_N)$. The column vector $\theta = (\theta_1, ..., \theta_J)^T$ is the generic parameter vector of mixture combining proportions. The likelihood function is

$$L_y(\theta) = \prod_{n=1}^{N} \Big(\theta_1 f_1(v_n) + ... + \theta_J f_J(v_n)\Big). \tag{13.44}$$

Then the log likelihood function is

$$LL_y(\theta) = \sum_{n=1}^{N} \log \Big(\theta_1 f_1(v_n) + ... + \theta_J f_J(v_n)\Big).$$

With u the column vector with entries $u_n = 1/N$, and P the matrix with entries $P_{nj} = f_j(v_n)$, we define

$$s_j = \sum_{n=1}^{N} P_{nj} = \sum_{n=1}^{N} f_j(v_n).$$

Maximizing $LL_y(\theta)$ is equivalent to minimizing

$$F(\theta) = KL(u, P\theta) + \sum_{j=1}^{J} (1 - s_j)\theta_j. \tag{13.45}$$

13.8.3 A Motivating Illustration

To motivate such mixture problems, we imagine that each data value is generated by first selecting one value of j, with probability θ_j, and then selecting a realization of a random variable governed by $f_j(v)$. For example, there could be J bowls of colored marbles, and we randomly select a bowl, and then randomly select a marble within the selected bowl. For each n the number v_n is the numerical code for the color of the nth marble drawn. In this illustration we are using a mixture of probability functions, but we could have used probability density functions.

13.8.4 The Acceptable Data

We approach the mixture problem by creating acceptable data. We imagine that we could have obtained $x_n = j_n$, for $n = 1, ..., N$, where the selection of v_n is governed by the function $f_{j_n}(v)$. In the bowls example, j_n is the number of the bowl from which the nth marble is drawn. The acceptable-data random vector is $X = (X_1, ..., X_N)$, where the X_n are independent random variables taking values in the set $\{j = 1, ..., J\}$. The value j_n is one realization of X_n. Since our objective is to estimate the true θ_j, the values v_n are now irrelevant. Our ML estimate of the true θ_j is simply the proportion of times $j = j_n$. Given a realization x of X, the conditional pdf or pf of Y does not involve the mixing proportions, so X is acceptable. Notice also that it is not possible to calculate the entries of y from those of x; the model $Y = h(X)$ does not hold.

13.8.5 The Mix-EM Algorithm

Using this acceptable data, we derive the EM algorithm, which we call the Mix-EM algorithm.

With N_j denoting the number of times the value j occurs as an entry of x, the likelihood function for X is

$$L_x(\theta) = f_X(x|\theta) = \prod_{j=1}^{J} \theta_j^{N_j}, \tag{13.46}$$

and the log likelihood is

$$LL_x(\theta) = \log L_x(\theta) = \sum_{j=1}^{J} N_j \log \theta_j. \tag{13.47}$$

Then

$$E(\log L_x(\theta)|y, \theta^k) = \sum_{j=1}^{J} E(N_j|y, \theta^k) \log \theta_j. \tag{13.48}$$

To simplify the calculations in the E-step we rewrite $LL_x(\theta)$ as

$$LL_x(\theta) = \sum_{n=1}^{N} \sum_{j=1}^{J} X_{nj} \log \theta_j, \tag{13.49}$$

where $X_{nj} = 1$ if $j = j_n$ and zero otherwise. Then we have

$$E(X_{nj}|y, \theta^k) = \text{prob}\,(X_{nj} = 1|y, \theta^k) = \frac{\theta_j^k f_j(v_n)}{f(v_n|\theta^k)}. \tag{13.50}$$

The function $E(LL_x(\theta)|y, \theta^k)$ becomes

$$E(LL_x(\theta)|y, \theta^k) = \sum_{n=1}^{N} \sum_{j=1}^{J} \frac{\theta_j^k f_j(v_n)}{f(v_n|\theta^k)} \log \theta_j. \tag{13.51}$$

Maximizing with respect to θ, we get the iterative step of the Mix-EM algorithm:

$$\theta_j^{k+1} = \frac{1}{N} \theta_j^k \sum_{n=1}^{N} \frac{f_j(v_n)}{f(v_n|\theta^k)}. \tag{13.52}$$

We know from our previous discussions that, since the preferred data X is acceptable, likelihood is non-decreasing for this algorithm. We shall go further now, and show that the sequence of probability vectors $\{\theta^k\}$ converges to a maximizer of the likelihood.

13.8.6 Convergence of the Mix-EM Algorithm

As we noted earlier, maximizing the likelihood in the mixture case is equivalent to minimizing

$$F(\theta) = KL(u, P\theta) + \sum_{j=1}^{J} (1 - s_j)\theta_j,$$

over probability vectors θ. It is easily shown that, if $\hat{\theta}$ minimizes $F(\theta)$ over all nonnegative vectors θ, then $\hat{\theta}$ is a probability vector. Therefore, we can obtain the maximum likelihood estimate of θ by minimizing $F(\theta)$ over nonnegative vectors θ.

The following theorem is found in [56].

Theorem 13.1 *Let u be any positive vector, P any nonnegative matrix with $s_j > 0$ for each j, and*

$$F(\theta) = KL(u, P\theta) + \sum_{j=1}^{J} \beta_j KL(\gamma_j, \theta_j).$$

If $s_j + \beta_j > 0$, $\alpha_j = s_j/(s_j + \beta_j)$, and $\beta_j \gamma_j \geq 0$, for all j, then the iterative sequence given by

$$\theta_j^{k+1} = \alpha_j s_j^{-1} \theta_j^k \Big(\sum_{n=1}^{N} P_{n,j} \frac{u_n}{(P\theta^k)_n} \Big) + (1 - \alpha_j)\gamma_j \tag{13.53}$$

converges to a nonnegative minimizer of $F(\theta)$.

With the choices $u_n = 1/N$, $\gamma_j = 0$, and $\beta_j = 1 - s_j$, the iteration in Equation (13.53) becomes that of the Mix-EM algorithm. Therefore, the sequence $\{\theta^k\}$ converges to the maximum likelihood estimate of the mixing proportions.

Chapter 14

Geometric Programming and the MART

14.1 Overview

Geometric Programming (GP) involves the minimization of functions of a special type, known as posynomials. The first systematic treatment of geometric programming appeared in the book [113], by Duffin, Peterson and Zener, the founders of geometric programming. As we shall see, the Generalized Arithmetic-Geometric Mean Inequality plays an important role in the theoretical treatment of geometric programming, particularly in the development of the dual GP (DGP) problem. The MART is then used to solve the DGP.

Although geometric programming is a fairly specialized topic, a detailed discussion of the GP problem is quite helpful in revealing new uses of familiar topics such as the Arithmetic-Geometric Mean Inequality, while introducing new themes, such as duality, primal and dual problems, and iterative computation, that play important roles in iterative optimization.

14.2 An Example of a GP Problem

The following optimization problem was presented originally by Duffin, *et al.* [113] and discussed by Peressini *et al.* in [186]. It illustrates well the type of problem considered in geometric programming. Suppose that 400 cubic yards of gravel must be ferried across a river in an open box of length t_1, width t_2 and height t_3. Each round-trip costs ten cents. The sides and the bottom of the box cost 10 dollars per square yard to build, while the ends of the box cost twenty dollars per square yard. The box will have no salvage value after it has been used. Determine the dimensions of the box that minimize the total cost.

With $t = (t_1, t_2, t_3)$, the cost function is

$$g(t) = \frac{40}{t_1 t_2 t_3} + 20 t_1 t_3 + 10 t_1 t_2 + 40 t_2 t_3, \tag{14.1}$$

which is to be minimized over $t_j > 0$, for $j = 1, 2, 3$. The function $g(t)$ is an example of a posynomial.

14.3 The Generalized AGM Inequality

The generalized arithmetic-geometric mean inequality will play a prominent role in solving the GP problem.

Suppose that $x_1, ..., x_N$ are positive numbers. Let $a_1, ..., a_N$ be positive numbers that sum to one. Then the *Generalized AGM Inequality* (GAGM Inequality) is

$$x_1^{a_1} x_2^{a_2} \cdots x_N^{a_N} \leq a_1 x_1 + a_2 x_2 + ... + a_N x_N, \tag{14.2}$$

with equality if and only if $x_1 = x_2 = ... = x_N$. We can prove this using the convexity of the function $-\log x$.

A function $f(x)$ is said to be *convex* over an interval (a, b) if

$$f(a_1 t_1 + a_2 t_2 + ... + a_N t_N) \leq a_1 f(t_1) + a_2 f(t_2) + ... + a_N f(t_N),$$

for all positive integers N, all a_n as above, and all real numbers t_n in (a, b). If the function $f(x)$ is twice differentiable on (a, b), then $f(x)$ is convex over (a, b) if and only if the second derivative of $f(x)$ is nonnegative on (a, b). For example, the function $f(x) = -\log x$ is convex on the positive x-axis. The GAGM Inequality follows immediately.

14.4 Posynomials and the GP Problem

Functions $g(t)$ of the form

$$g(t) = \sum_{i=1}^{n} c_i \Big(\prod_{j=1}^{m} t_j^{a_{ij}} \Big), \tag{14.3}$$

with $t = (t_1, ..., t_m)$, the $t_j > 0$, $c_i > 0$ and a_{ij} real, are called *posynomials.* The *geometric programming problem*, denoted (GP), is to minimize a given posynomial over positive t. In order for the minimum to be greater than zero, we need some of the a_{ij} to be negative.

We denote by $u_i(t)$ the function

$$u_i(t) = c_i \prod_{j=1}^{m} t_j^{a_{ij}}, \tag{14.4}$$

so that

$$g(t) = \sum_{i=1}^{n} u_i(t). \tag{14.5}$$

For any choice of $\delta_i > 0$, $i = 1, ..., n$, with

$$\sum_{i=1}^{n} \delta_i = 1,$$

we have

$$g(t) = \sum_{i=1}^{n} \delta_i \Big(\frac{u_i(t)}{\delta_i} \Big). \tag{14.6}$$

Applying the Generalized Arithmetic-Geometric Mean (GAGM) Inequality, we have

$$g(t) \geq \prod_{i=1}^{n} \Big(\frac{u_i(t)}{\delta_i} \Big)^{\delta_i}. \tag{14.7}$$

Therefore,

$$g(t) \geq \prod_{i=1}^{n} \Big(\frac{c_i}{\delta_i} \Big)^{\delta_i} \left(\prod_{i=1}^{n} \prod_{j=1}^{m} t_j^{a_{ij}\delta_i} \right), \tag{14.8}$$

or

$$g(t) \geq \prod_{i=1}^{n} \left(\frac{c_i}{\delta_i}\right)^{\delta_i} \left(\prod_{j=1}^{m} t_j^{\sum_{i=1}^{n} a_{ij}\delta_i}\right). \tag{14.9}$$

Suppose that we can find $\delta_i > 0$ with

$$\sum_{i=1}^{n} a_{ij}\delta_i = 0, \tag{14.10}$$

for each j. Then the inequality in (14.9) becomes

$$g(t) \geq v(\delta), \tag{14.11}$$

for

$$v(\delta) = \prod_{i=1}^{n} \left(\frac{c_i}{\delta_i}\right)^{\delta_i}. \tag{14.12}$$

14.5 The Dual GP Problem

The *dual geometric programming problem*, denoted (DGP), is to maximize the function $v(\delta)$, over all *feasible* $\delta = (\delta_1, ..., \delta_n)$, that is, all positive δ for which

$$\sum_{i=1}^{n} \delta_i = 1, \tag{14.13}$$

and

$$\sum_{i=1}^{n} a_{ij}\delta_i = 0, \tag{14.14}$$

for each $j = 1, ..., m$. Clearly, we have

$$g(t) \geq v(\delta), \tag{14.15}$$

for any positive t and feasible δ. Of course, there may be no feasible δ, in which case (DGP) is said to be *inconsistent*.

As we have seen, the inequality in (14.15) is based on the GAGM Inequality. We have equality in the GAGM Inequality if and only if the terms in the arithmetic mean are all equal. In this case, this says that there is a constant λ such that

$$\frac{u_i(t)}{\delta_i} = \lambda, \tag{14.16}$$

for each $i = 1, ..., n$. Using the fact that the δ_i sum to one, it follows that

$$\lambda = \sum_{i-1}^{n} u_i(t) = g(t), \tag{14.17}$$

and

$$\delta_i = \frac{u_i(t)}{g(t)}, \tag{14.18}$$

for each $i = 1, ..., n$. As the theorem below asserts, if t^* is positive and minimizes $g(t)$, then δ^*, the associated δ from Equation (14.18), is feasible and solves (DGP). Since we have equality in the GAGM Inequality now, we have

$$g(t^*) = v(\delta^*).$$

The main theorem in geometric programming is the following.

Theorem 14.1 *If $t^* > 0$ minimizes $g(t)$, then (DGP) is consistent. In addition, the choice*

$$\delta_i^* = \frac{u_i(t^*)}{g(t^*)} \tag{14.19}$$

is feasible and solves (DGP). Finally,

$$g(t^*) - v(\delta^*); \tag{14.20}$$

that is, there is no duality gap.

Proof: We have

$$\frac{\partial u_i}{\partial t_j}(t^*) = \frac{a_{ij} u_i(t^*)}{t_j^*}, \tag{14.21}$$

so that

$$t_j^* \frac{\partial u_i}{\partial t_j}(t^*) = a_{ij} u_i(t^*), \tag{14.22}$$

for each $j = 1, ..., m$. Since t^* minimizes $g(t)$, we have

$$0 = \frac{\partial g}{\partial t_j}(t^*) = \sum_{i=1}^{n} \frac{\partial u_i}{\partial t_j}(t^*), \tag{14.23}$$

so that, from Equation (14.22), we have

$$0 = \sum_{i=1}^{n} a_{ij} u_i(t^*), \tag{14.24}$$

for each $j = 1, ..., m$. It follows that δ^* is feasible. Since we have equality in the GAGM Inequality, we know

$$g(t^*) = v(\delta^*). \tag{14.25}$$

Therefore, δ^* solves (DGP). This completes the proof. ∎

14.6 Solving the GP Problem

The theorem suggests how we might go about solving (GP). First, we try to find a feasible δ^* that maximizes $v(\delta)$. This means we have to find a positive solution to the system of $m+1$ linear equations in n unknowns, given by

$$\sum_{i=1}^{n} \delta_i = 1, \tag{14.26}$$

and

$$\sum_{i=1}^{n} a_{ij} \delta_i = 0, \tag{14.27}$$

for $j = 1, ..., m$, such that $v(\delta)$ is maximized. As we shall see, the *multiplicative algebraic reconstruction technique* (MART) is an iterative procedure that we can use to find such δ. If there is no such vector, then (GP) has no minimizer. Once the desired δ^* has been found, we set

$$\delta_i^* = \frac{u_i(t^*)}{v(\delta^*)}, \tag{14.28}$$

for each $i = 1, ..., n$, and then solve for the entries of t^*. This last step can be simplified by taking logs; then we have a system of linear equations to solve for the values $\log t_j^*$.

14.7 Solving the DGP Problem

The iterative multiplicative algebraic reconstruction technique MART can be used to minimize the function $v(\delta)$, subject to linear equality constraints, provided that the matrix involved has nonnegative entries. We cannot apply the MART yet, because the matrix A^T does not satisfy these conditions.

14.7.1 The MART

The MART is an iterative algorithm for finding a nonnegative solution of the system $Px = y$, for an I by J matrix P with nonnegative entries and vector y with positive entries. We also assume that

$$p_j = \sum_{i=1}^{I} P_{ij} > 0,$$

for all $i = 1, ..., I$. When discussing the MART, we say that the system $Px = y$ is *consistent* when it has nonnegative solutions. We consider two different versions of the MART.

14.7.1.1 MART I

The iterative step of the first version of MART, which we shall call MART I, is the following: for $k = 0, 1, ...,$ and $i = k(\text{mod}\, I) + 1$, let

$$x_j^{k+1} = x_j^k \Big(\frac{y_i}{(Px^k)_i}\Big)^{P_{ij}/m_i},$$

for $j = 1, ..., J$, where the parameter m_i is defined to be

$$m_i = \max\{P_{ij} | j = 1, ..., J\}.$$

The MART I algorithm converges, in the consistent case, to the nonnegative solution for which the KL distance $KL(x, x^0)$ is minimized.

14.7.1.2 MART II

The iterative step of the second version of MART, which we shall call MART II, is the following: for $k = 0, 1, ...,$ and $i = k(\text{mod}\, I) + 1$, let

$$x_j^{k+1} = x_j^k \Big(\frac{y_i}{(Px^k)_i}\Big)^{P_{ij}/p_j n_i},$$

for $j = 1, ..., J$, where the parameter n_i is defined to be

$$n_i = \max\{P_{ij}p_j^{-1}|j = 1, ..., J\}.$$

The MART II algorithm converges, in the consistent case, to the nonnegative solution for which the KL distance

$$\sum_{j=1}^{J} p_j KL(x_j, x_j^0)$$

is minimized.

14.7.2 Using the MART to Solve the DGP Problem

Let the $(n+1)$ by m matrix A^T have the entries $A_{ji} = a_{ij}$, for $j = 1, ..., m$ and $i = 1, ..., n$, and $A_{(m+1),i} = 1$. Let u be the column vector with entries $u_j = 0$, for $j = 1, ..., m$, and $u_{m+1} = 1$.

The entries on the bottom row of A^T are all one, as is the bottom entry of the column vector u, since these entries correspond to the equation $\sum_{i=1}^{I} \delta_i = 1$. By adding suitably large positive multiples of this last equation to the other equations in the system, we obtain an equivalent system, $B^T\delta = s$, for which the new matrix B^T and the new vector s have only positive entries. Now we can apply the MART I algorithm to the system $B^T\delta = s$, letting $P = B^T$, $p_i = \sum_{j=1}^{J+1} B_{ij}$, $\delta = x$, $x^0 = c$ and $y = s$. In the consistent case, the MART I algorithm will find the nonnegative solution that minimizes $KL(x, x^0)$, so we select $x^0 = c$. Then the MART I algorithm finds the nonnegative δ^* satisfying $B^T\delta^* = s$, or, equivalently, $A^T\delta^* = u$, for which the KL distance

$$KL(\delta, c) = \sum_{i=1}^{I} \left(\delta_i \log \frac{\delta_i}{c_i} + c_i - \delta_i\right)$$

is minimized. Since we know that

$$\sum_{i=1}^{I} \delta_i = 1,$$

it follows that minimizing $KL(\delta, c)$ is equivalent to maximizing $v(\delta)$. Using δ^*, we find the optimal t^* solving the GP problem.

For example, the linear system of equations $A^T\delta = u$ corresponding to the posynomial in Equation (14.1) is

$$A^T\delta = u = \begin{bmatrix} -1 & 1 & 1 & 0 \\ -1 & 0 & 1 & 1 \\ -1 & 1 & 0 & 1 \\ 1 & 1 & 1 & 1 \end{bmatrix} \begin{bmatrix} \delta_1 \\ \delta_2 \\ \delta_3 \\ \delta_4 \end{bmatrix} = \begin{bmatrix} 0 \\ 0 \\ 0 \\ 1 \end{bmatrix}.$$

Adding two times the last row to the other rows, the system becomes

$$B^T\delta = s = \begin{bmatrix} 1 & 3 & 3 & 2 \\ 1 & 2 & 3 & 3 \\ 1 & 3 & 2 & 3 \\ 1 & 1 & 1 & 1 \end{bmatrix} \begin{bmatrix} \delta_1 \\ \delta_2 \\ \delta_3 \\ \delta_4 \end{bmatrix} = \begin{bmatrix} 2 \\ 2 \\ 2 \\ 1 \end{bmatrix}.$$

The matrix B^T and the vector s are now positive. We are ready to apply the MART.

The MART iteration is as follows. With $j = k(\mathrm{mod}\,(J+1)) + 1$, $m_j = \max\{B_{ij}\,|i = 1, 2, ..., I\}$ and $k = 0, 1, ...$, let

$$\delta_i^{k+1} = \delta_i^k \left(\frac{s_j}{(B^T\delta^k)_j} \right)^{m_j^{-1} B_{ij}}.$$

The optimal δ^* is $\delta^* = (.4, .2, .2, .2)^T$, the optimal t^* is $t^* = (2, 1, .5)$, and the lowest cost is one hundred dollars.

14.8 Constrained Geometric Programming

Consider now the following variant of the problem of transporting the gravel across the river. Suppose that the bottom and the two sides will be constructed for free from scrap metal, but only four square yards are available. The cost function to be minimized becomes

$$g_0(t) = \frac{40}{t_1t_2t_3} + 40t_2t_3, \tag{14.29}$$

and the constraint is

$$g_1(t) = \frac{t_1t_3}{2} + \frac{t_1t_2}{4} \leq 1. \tag{14.30}$$

With $\delta_1 > 0$, $\delta_2 > 0$, and $\delta_1 + \delta_2 = 1$, we write

$$g_0(t) = \delta_1 \frac{40}{\delta_1 t_1t_2t_3} + \delta_2 \frac{40t_2t_3}{\delta_2}. \tag{14.31}$$

Since $0 \leq g_1(t) \leq 1$, we have

$$g_0(t) \geq \left(\delta_1 \frac{40}{\delta_1 t_1t_2t_3} + \delta_2 \frac{40t_2t_3}{\delta_2}\right)\left(g_1(t)\right)^{\lambda}, \tag{14.32}$$

for any positive λ. The GAGM Inequality then tells us that

$$g_0(t) \geq \left(\left(\frac{40}{\delta_1 t_1t_2t_3}\right)^{\delta_1} \left(\frac{40t_2t_3}{\delta_2}\right)^{\delta_2}\right)\left(g_1(t)\right)^{\lambda}, \tag{14.33}$$

so that

$$g_0(t) \geq \left(\left(\frac{40}{\delta_1}\right)^{\delta_1}\left(\frac{40}{\delta_2}\right)^{\delta_2}\right) t_1^{-\delta_1} t_2^{\delta_2-\delta_1} t_3^{\delta_2-\delta_1}\Big(g_1(t)\Big)^{\lambda}. \tag{14.34}$$

From the GAGM Inequality, we also know that, for $\delta_3 > 0$, $\delta_4 > 0$ and $\lambda = \delta_3 + \delta_4$,

$$\Big(g_1(t)\Big)^{\lambda} \geq (\lambda)^{\lambda}\left(\left(\frac{1}{2\delta_3}\right)^{\delta_3}\left(\frac{1}{4\delta_4}\right)^{\delta_4}\right) t_1^{\delta_3+\delta_4} t_2^{\delta_4} t_3^{\delta_3}. \tag{14.35}$$

Combining the inequalities in (14.34) and (14.35), we obtain

$$g_0(t) \geq v(\delta) t_1^{-\delta_1+\delta_3+\delta_4} t_2^{-\delta_1+\delta_2+\delta_4} t_3^{-\delta_1+\delta_2+\delta_3}, \tag{14.36}$$

with

$$v(\delta) = \left(\frac{40}{\delta_1}\right)^{\delta_1}\left(\frac{40}{\delta_2}\right)^{\delta_2}\left(\frac{1}{2\delta_3}\right)^{\delta_3}\left(\frac{1}{4\delta_4}\right)^{\delta_4}\Big(\delta_3+\delta_4\Big)^{\delta_3+\delta_4}, \tag{14.37}$$

and $\delta = (\delta_1, \delta_2, \delta_3, \delta_4)$. If we can find a positive vector δ with

$$\begin{aligned} \delta_1 + \delta_2 &= 1, \\ \delta_3 + \delta_4 &= \lambda, \\ -\delta_1 + \delta_3 + \delta_4 &= 0, \\ -\delta_1 + \delta_2 + \delta_4 &= 0 \\ -\delta_1 + \delta_2 + \delta_3 &= 0, \end{aligned} \tag{14.38}$$

then

$$g_0(t) \geq v(\delta). \tag{14.39}$$

In this particular case, there is a unique positive δ satisfying the equations (14.38), namely

$$\delta_1^* = \frac{2}{3}, \delta_2^* = \frac{1}{3}, \delta_3^* = \frac{1}{3}, \text{and } \delta_4^* = \frac{1}{3}, \tag{14.40}$$

and

$$v(\delta^*) = 60. \tag{14.41}$$

Therefore, $g_0(t)$ is bounded below by 60. If there is t^* such that

$$g_0(t^*) = 60, \tag{14.42}$$

then we must have

$$g_1(t^*) = 1, \tag{14.43}$$

and equality in the GAGM Inequality. Consequently,

$$\frac{3}{2}\frac{40}{t_1^* t_2^* t_3^*} = 3(40 t_2^* t_3^*) = 60, \tag{14.44}$$

and

$$\frac{3}{2} t_1^* t_3^* = \frac{3}{4} t_1^* t_2^* = K. \tag{14.45}$$

Since $g_1(t^*) = 1$, we must have $K = \frac{3}{2}$. We solve these equations by taking logarithms, to obtain the solution

$$t_1^* = 2,\, t_2^* = 1,\, \text{and}\, t_3^* = \frac{1}{2}. \tag{14.46}$$

The change of variables $t_j = e^{x_j}$ converts the constrained (GP) problem into a constrained convex programming problem. The theory of the constrained (GP) problem can then be obtained as a consequence of the theory for the convex programming problem.

See [28] for a discussion of the use of constrained GP to find the Perron-Frobenius eigenvalue of a positive matrix.

14.9 Exercises

Ex. 14.1 *Show that there is no solution to the problem of minimizing the function*

$$g(t_1, t_2) = \frac{2}{t_1 t_2} + t_1 t_2 + t_1, \tag{14.47}$$

over $t_1 > 0$, $t_2 > 0$.

Ex. 14.2 *Minimize the function*

$$g(t_1, t_2) = \frac{1}{t_1 t_2} + t_1 t_2 + t_1 + t_2, \tag{14.48}$$

over $t_1 > 0$, $t_2 > 0$. *This will require some iterative numerical method for solving equations.*

Ex. 14.3 *Program the MART algorithm and use it to verify the assertions made previously concerning the solutions of the two numerical examples.*

Chapter 15

Variational Inequality Problems and Algorithms

15.1 Monotone Functions

Variational inequality problems (VIP) generalize the problem of minimizing a convex function over a closed convex set. Saddle-point problems can be reformulated as variational inequality problems (VIP) for monotone functions, and iterative algorithms used for their solution. Throughout this chapter the norm is the Euclidean norm. We begin with some definitions.

A function $f : \mathbb{R}^J \to [-\infty, +\infty]$ is *proper* if there is no x with $f(x) = -\infty$ and some x with $f(x) < +\infty$. The *effective domain* of f, denoted $\text{dom}(f)$, is the set of all x for which $f(x)$ is finite. If f is a proper convex function on $\mathbb{R}^J$, then the sub-differential $\partial f(x)$, defined to be the set

$$\partial f(x) = \{u | f(z) \geq f(x) + \langle u, z - x\rangle \,\text{for all}\, z\}, \tag{15.1}$$

is a closed convex set, and nonempty for every x in the interior of $\text{dom}(f)$.

This is a consequence of applying the Support Theorem to the epi-graph of f. We say that f is *differentiable* at x if $\partial f(x)$ is a singleton set, in which case we have $\partial f(x) = \{\nabla f(x)\}$.

Definition 15.1 *An operator* $T : \mathbb{R}^J \to \mathbb{R}^J$ *is* monotone *if*

$$\langle Tx - Ty, x - y\rangle \geq 0,$$

for all x *and* y.

Definition 15.2 *An operator* $T : \mathbb{R}^J \to \mathbb{R}^J$ *is* strongly monotone *if*

$$\langle Tx - Ty, x - y\rangle \geq \nu\|x - y\|^2,$$

for all x *and* y.

As we saw previously, an operator $G : \mathbb{R}^J \to \mathbb{R}^J$ is ν-inverse strongly monotone (ν-ism) if

$$\langle Gx - Gy, x - y\rangle \geq \nu\|Gx - Gy\|^2,$$

for all x and y.

Let $f : \mathbb{R}^J \to \mathbb{R}$ be convex and differentiable. Then the operator $T = \nabla f$ is monotone. If $f(x)$ is convex, but not differentiable, then $B(x) = \partial f(x)$ is a monotone set-valued function; we shall discuss set-valued functions in Chapter 16. Not all monotone operators are gradient operators, as Exercise 15.1 will show. In fact, if A is a non-zero, skew-symmetric matrix, then $Tx = Ax$ is a monotone operator, but is not a gradient operator.

It is easy to see that if N is ne, then $I - N$ is monotone.

Definition 15.3 *An operator* $G : \mathbb{R}^J \to \mathbb{R}^J$ *is* weakly ν-inverse strongly monotone *if*

$$\langle Gx, x - y\rangle \geq \nu\|Gx\|^2, \tag{15.2}$$

whenever $Gy = 0$.

15.2 The Split-Feasibility Problem

The split-feasibility problem (SFP) is the following: find x in C with Ax in Q, where A is an I by J matrix, and C and Q nonempty, closed convex sets in $\mathbb{R}^J$ and $\mathbb{R}^I$, respectively. The CQ algorithm [58, 59] has the iterative step

$$x^{k+1} = P_C(I - \gamma A^T(I - P_Q)A)x^k. \tag{15.3}$$

For $0 < \gamma < \frac{2}{\rho(A^TA)}$, the sequence $\{x^k\}$ converges to a minimizer, over x in C, of the convex function

$$f(x) = \frac{1}{2}\|P_QAx - Ax\|^2,$$

whenever such minimizers exist. From Theorem 8.1 we know that the gradient of $f(x)$ is

$$\nabla f(x) = A^T(I - P_Q)Ax,$$

so the iteration in Equation (15.3) can be written as

$$x^{k+1} = P_C(I - \gamma\nabla f)x^k. \tag{15.4}$$

The limit x^* of the sequence $\{x^k\}$ satisfies the inequality

$$\langle\nabla f(x^*), c - x^*\rangle \geq 0, \tag{15.5}$$

for all c in C.

15.3 The Variational Inequality Problem

Now let G be any operator on $\mathbb{R}^J$. The *variational inequality problem* (VIP), with respect to G and C, denoted VIP(G, C), is to find an x^* in C such that

$$\langle Gx^*, c - x^*\rangle \geq 0,$$

for all c in C. The form of the CQ algorithm suggests that we consider solving the VIP(G, C) using the following iterative scheme:

$$x^{k+1} = P_C(I - \gamma G)x^k. \tag{15.6}$$

The sequence $\{x^k\}$ solves the VIP(G, C) whenever there are solutions, if G is ν-ism and $0 < \gamma < 2\nu$; this is sometimes called Dolidze's Theorem, and is proven in [59] (see also [133]). A good source for related algorithms is the paper by Censor, Iusem and Zenios [90].

In [90] the authors mention that it has been shown that, if G is strongly monotone and L-Lipschitz, then the iteration in Equation (15.6) converges to a solution of VIP(G, C) whenever $\gamma \in (0, 2\nu/L^2)$. Then we have a strict contraction mapping, a fixed point necessarily exists, and the result follows. But under these conditions, $I-\gamma G$ is also averaged, so the result, except for the existence of fixed points, follows from Dolidze's Theorem. When G is not ism, there are other iterative algorithms that can be used; for example, Korpelevich's algorithm [151] has been studied extensively and discussed in detail in [120]. We discuss this method in the next section.

15.4 Korpelevich's Method for the VIP

An operator T on $\mathbb{R}^J$ is *pseudo-monotone* if

$$\langle Ty, x-y\rangle \geq 0$$

implies

$$\langle Tx, x-y\rangle \geq 0.$$

Any monotone operator is pseudo-monotone.

Suppose now that G is L-Lipschitz and pseudo-monotone, but not necessarily ism. Let $\gamma L < 1$, and $S = \gamma G$. Korpelevich's algorithm is then

$$x^{k+1} = P_C(x^k - Sy^k), \tag{15.7}$$

where

$$y^k = P_C(x^k - Sx^k). \tag{15.8}$$

The sequence $\{x^k\}$ converges to a solution of VIP(G, C) whenever there are solutions [151, 86].

15.4.1 The Special Case of $C = \mathbb{R}^J$

In the special case of the VIP(G, C) in which $C = \mathbb{R}^J$ and $P_C = I$, Korpelevich's algorithm employs the iterative steps

$$x^{k+1} = x^k - Sy^k, \tag{15.9}$$

where

$$y^k = x^k - Sx^k. \tag{15.10}$$

Then we have

$$x^{k+1} = (I - S(I-S))x^k. \tag{15.11}$$

If the operator $S(I-S)$ is ν-ism for some $\nu > \frac{1}{2}$, then the sequence $\{x^k\}$ converges to a solution of VIP$(G, \mathbb{R}^J)$ whenever there are solutions, according to the KMO Theorem. Note that z solves the VIP$(G, \mathbb{R}^J)$ if and only if $0 \in \partial G(z)$. The KMO Theorem is valid whenever G is weakly ν-ism for some $\nu > \frac{1}{2}$. Therefore, we get convergence of this special case of the Korpelevich iteration by showing that the operator $S(I-S)$ is weakly $\frac{1}{1+\sigma}$-ism, where $\sigma = \gamma L < 1$.

Our assumptions are that $T = I - S(I-S)$, S is pseudo-monotone and σ-Lipschitz, for some $\sigma < 1$, and $Tz = z$. It follows then that $S(I-S)z = 0$. From

$$\|Sz\| = \|Sz - S(I-S)z\| \leq \sigma\|z - (I-S)z\| = \sigma\|Sz\|,$$

and the fact that $\sigma < 1$, we conclude that $Sz = 0$ as well.

Lemma 15.1 *Let x be arbitrary, and $z = Tz$. Then*

$$2\langle S(I-S)x, x-z\rangle \geq (1-\sigma^2)\|Sx\|^2 + \|S(I-S)x\|^2. \tag{15.12}$$

Proof: Using $Sz = S(I-S)z = 0$, we write

$$2\langle S(I-S)x, x-z\rangle = 2\langle S(I-S)x - S(I-S)z, x - Sx - z + Sz\rangle + 2\langle S(I-S)x, Sx\rangle$$
$$= 2\langle S(I-S)x - S(I-S)z, (I-S)x - (I-S)z\rangle + 2\langle S(I-S)x, Sx\rangle$$
$$\geq 2\langle S(I-S)x, Sx\rangle.$$

Also, we have

$$\|S(I-S)x\|^2 - 2\langle S(I-S)x, Sx\rangle + \|Sx\|^2 = \|S(I-S)x - Sx\|^2 \leq \sigma^2\|Sx\|^2.$$

Therefore,

$$2\langle S(I-S)x, x-z\rangle \geq 2\langle S(I-S)x, Sx\rangle \geq (1-\sigma^2)\|Sx\|^2 + \|S(I-S)\|^2.$$

∎

It follows from (15.12) and Cauchy's Inequality that

$$2\|Sx\|\|S(I-S)x\| \geq 2\langle S(I-S)x, Sx\rangle \geq (1-\sigma^2)\|Sx\|^2 + \|S(I-S)\|^2,$$

so that

$$\sigma^2\|Sx\|^2 \geq (\|Sx\| - \|S(I-S)x\|)^2.$$

Therefore,

$$(1+\sigma)\|Sx\| \geq \|S(I-S)x\|.$$

From

$$\langle S(I-S)x, x-z\rangle \geq \frac{1-\sigma^2}{2}\|Sx\|^2 + \frac{1}{2}\|S(I-S)\|^2$$

and

$$\|Sx\|^2 > \frac{1}{(1+\sigma)^2}\|S(I-S)x\|^2$$

we get

$$\langle S(I-S)x, x-z\rangle \geq \frac{1}{1+\sigma}\|S(I-S)x\|^2;$$

in other words, the operator $S(I-S)$ is weakly $\frac{1}{1+\sigma}$-ism.

15.4.2 The General Case

Now we have $x^{k+1} = Tx^k$ where $S = \gamma G$ and $T = P_C(I - SP_C(I - S))$. We assume that $Tz = z$, so that z solves VIP(G, C).

The key Proposition now is the following.

Proposition 15.1 *Let $G : C \to \mathbb{R}^J$ be pseudo-monotone and L-Lipschitz, let $\sigma = \gamma L < 1$, and let $S = \gamma G$. For any k let $y^k = P_C(I - S)x^k$. Then*

$$\|z - x^k\|^2 - \|z - x^{k+1}\|^2 \geq (1 - \sigma^2)\|y^k - x^k\|^2. \tag{15.13}$$

The proof of Proposition 15.1 follows that in [120]. The inequality in (15.13) emerges as a consequence of a sequence of inequalities and equations. We list these results first, and then discuss their proofs. For convenience, we let $w^k = x^k - Sx^k$.

1. $\langle Sy^k, y^k - x^{k+1}\rangle \geq \langle Sy^k, z - x^{k+1}\rangle$.

2. $\langle Sx^k - Sy^k, x^{k+1} - y^k\rangle \geq \langle w^k - y^k, x^{k+1} - y^k\rangle$.

3. $\|z - w^k|^2 - \|w^k - x^{k+1}\|^2 \geq \|z - x^{k+1}\|^2$.

4. $\|z - w^k\|^2 - \|w^k - x^{k+1}\|^2 =$
$$\|z - x^k\|^2 - \|x^k - x^{k+1}\|^2 + 2\langle Sy^k, z - x^{k+1}\rangle.$$

5. $\|z - w^k\|^2 - \|w^k - x^{k+1}\|^2 \leq$
$$\|z - x^k\|^2 - \|x^k - x^{k+1}\|^2 + 2\langle Sy^k, y^k - x^{k+1}\rangle.$$

6. $-\|x^k - x^{k+1}\|^2 + 2\langle Sy^k, y^k - x^{k+1}\rangle =$
$$-\|y^k - x^k\|^2 - \|y^k - x^{k+1}\|^2 + 2\langle w^k - y^k, x^{k+1} - y^k\rangle.$$

7. $\|z - x^k\|^2 - \|z - x^{k+1}\|^2 \geq$
$$\|y^k - x^k\|^2 + \|y^k - x^{k+1}\|^2 - 2\langle y^k - w^k, y^k - x^{k+1}\rangle.$$

8. $2\langle y^k - w^k, y^k - x^{k+1}\rangle \leq 2\gamma L\|y^k - x^{k+1}\|\|y^k - x^k\|$.

9. $2\gamma L\|y^k - x^{k+1}\|\|y^k - x^k\| \leq \gamma^2 L^2\|y^k - x^k\|^2 + \|y^k - x^{k+1}\|^2$.

10. $2\gamma L\|y^k - x^{k+1}\|\|y^k - x^k\| \leq \gamma L(\|y^k - x^k\|^2 + \|y^k - x^{k+1}\|^2)$.

11. $\|z - x^k\|^2 - \|z - x^{k+1}\|^2 \geq (1 - \gamma^2 L^2)\|y^k - x^k\|^2$.

12. $\|z - x^k\|^2 - \|z - x^{k+1}\|^2 \geq (1 - \gamma L)(\|y^k - x^k\|^2 + \|y^k - x^{k+1}\|^2)$.

Inequality 1 follows from the fact that z solves the VIP(G, C) and is pseudo-monotone, and x^{k+1} is in C. To obtain Inequality 2, add and subtract Sx^k on the left side of the inner product in 1 and use the fact that $y^k = P_C(I - S)x^k$.

To get Inequality 3, expand

$$\|z - x^{k+1}\|^2 = \|z - w^k + w^k - x^{k+1}\|^2,$$

add and subtract x^{k+1} and use the fact that $x^{k+1} = P_C w^k$.

To get Equation 4 use $w^k = x^k - Sy^k$ and expand. Then Inequality 5 follows from 4 using Inequality 1. Equation 6 is easy.

Inequality 7 follows from 3, 5 and 6. To get Inequality 8, use $w^k = x^k - Sy^k$, add and subtract Sx^k in the left side of the inner product, and use $y^k = P_C(I - S)x^k$.

To get Inequality 9, expand

$$(\gamma L\|y^k - x^k\| - \|y^k - x^{k+1}\|)^2.$$

Then 11 is immediate, and the Proposition is proved. To get Inequality 10, expand

$$(\|y^k - x^k\| - \|y^k - x^{k+1}\|)^2.$$

Then 12 is immediate. We shall use 12 in a moment.

From Inequality (15.13), we learn that the sequence $\{\|z - x^k\|\}$ is decreasing, and so the sequence $\|y^k - x^k\|\}$ converges to zero. From 12 we learn that the sequence $\{\|y^k - x^{k+1}\|\}$ converges to zero.

We know that

$$\|x^k - x^{k+1}\|^2 = \|x^k - y^k + y^k - x^{k+1}\|^2 =$$

$$\|x^k - y^k\|^2 + \|y^k - x^{k+1}\|^2 + 2\langle x^k - y^k, y^k - x^{k+1}\rangle \leq$$

$$\|x^k - y^k\|^2 + \|y^k - x^{k+1}\|^2 + 2\|x^k - y^k\|\|x^{k+1} - y^k\|,$$

so it follows that $\{\|x^k - x^{k+1}\|\}$ converges to zero. The sequence $\{x^k\}$ is bounded; let x^* be a cluster point. Then x^* is a fixed point; that is

$$x^* = P_C(x^* - S(I - S)x^*),$$

so x^* solves the VIP(G, C) and we can replace z with x^* in all the lines above. It follows that $\{x^k\}$ converges to x^*. Therefore, the Korpelevich iteration converges whenever there is a solution of the VIP(G, C).

We saw that in the special case of $P_C = I$, the operator $S(I - S)$ is weakly ν-ism; it does not appear to be true in the general case that weak ism plays a role.

15.5 On Some Algorithms of Noor

In this section I comment on two algorithms that appear in the papers of Noor.

15.5.1 A Conjecture

We saw that for the case of $C = \mathbb{R}^J$ the operator $T = I - S(I - S)$ generates a sequence that converges to a fixed point of T. I suspected that the operator $P = (I - S)^2 = T - S$ might also work. More generally, I conjectured that the operator

$$Px = P_C(P_C(x - Sx) - SP_C(x - Sx)) = (P_C(I - S))^2 x \qquad (15.14)$$

would work for the general case. Noor [178, 179, 180] considers this and related methods, but does not provide details for this particular algorithm. The conjecture is false, but we can at least show that $z = Pz$ if and only if z solves VIP(G, C).

Proposition 15.2 *We have $z = Pz$ if and only if z solves VIP(G, C).*

Proof: One way is clear. So assume that $z = Pz$. Let $y = P_C(z - Sz)$, so that $z = P_C(y - Sy)$. Then for all $c \in C$ we have

$$\langle z - y + Sy, c - z \rangle \geq 0,$$

and

$$\langle y - z + Sz, c - y \rangle \geq 0.$$

Therefore,

$$\langle Sy, y - z \rangle \geq \|y - z\|^2,$$

and

$$-\langle Sz, y - z \rangle \geq \|y - z\|^2.$$

Adding, we get

$$\sigma \|y - z\|^2 \geq \|Sy - Sz\| \, \|y - z\| \geq \langle Sy - Sz, y - z \rangle \geq 2\|y - z\|^2,$$

from which we conclude that $y = z$. ∎

Unfortunately, this algorithm, which is Algorithm 3.6 in Noor [179], fails to converge, in general, as the following example shows. Let S be the operator on $\mathbb{R}^2$ given by multiplication by the matrix

$$S = \begin{bmatrix} 0 & a \\ -a & 0 \end{bmatrix},$$

for some $a \in (0,1)$. The operator S is then monotone and a-Lipschitz continuous. With $C = \mathbb{R}^2$, the variational inequality problem is then equivalent to finding a zero of S. Note that $Sz = 0$ if and only if $z = 0$.

The Korpelevich iteration in this case is

$$x^{k+1} = Tx^k = (I - S(I - S))x^k.$$

Noor's Algorithm 3.6 now has the iterative step

$$x^{k+1} = Px^k - (I - S)^2 x^k.$$

The operator T is then multiplication by the matrix

$$T = \begin{bmatrix} 1 - a^2 & -a \\ a & 1 - a^2 \end{bmatrix},$$

and the operator P is multiplication by the matrix

$$P = \begin{bmatrix} 1 - a^2 & -2a \\ 2a & 1 - a^2 \end{bmatrix}.$$

For any $x \in \mathbb{R}^2$ we have

$$\|Tx\|^2 = ((1 - a^2)^2 + a^2)\|x\|^2 < \|x\|^2,$$

for all $x \neq 0$, while

$$\|Px\|^2 = ((1 - a^2)^2 + 4a^2)\|x\|^2 = (1 + a^2)^2\|x\|^2.$$

This proves that the sequence $x^{k+1} = Px^k$ does not converge, generally.

15.6 Split Variational Inequality Problems

The *split variational inequality* problem (SVIP) is the following: find x^* in C such that

$$\langle f(x^*), c - x^* \rangle \geq 0, \tag{15.15}$$

for all $c \in C$, and

$$\langle g(Ax^*), q - Ax^* \rangle \geq 0, \tag{15.16}$$

for all $q \in Q$.

In [87] the authors present an iterative algorithm for solving the SVIP (see also [172]). The iterative step is $x^{k+1} = Sx^k$, where

$$S = U(I + \gamma A^T(T - I)A),$$

$$U = P_C(I - \lambda f),$$

and

$$T = P_Q(I - \lambda g).$$

It is easy to show that x^* satisfies Equation (15.15) if and only if x^* is a fixed point of U, and Ax^* satisfies Equation (15.16) if and only if Ax^* is a fixed point of T. We have the following convergence theorem for the sequence $\{x^k\}$.

Theorem 15.1 *Let f be ν_1-ism, g be ν_2-ism, $\nu = \min\{\nu_1, \nu_2\}$, $\lambda \in (0, 2\alpha)$, and $\gamma \in (0, 1/L)$, where L is the spectral radius of A^TA. If the SVIP has solutions, then the sequence $\{x^k\}$ converges to a solution of the SVIP.*

Take $\lambda < 2\nu$. Then the operators $I - \lambda f$ and $I - \lambda g$ are δ-av, for $\delta = \frac{\lambda}{2\nu} < 1$. The operator P_Q is firmly non-expansive, so is $\frac{1}{2}$-av. Then the operator $T = P_Q(I - \lambda g)$ is ϕ-av, with $\phi = \frac{\delta+1}{2}$.

The following lemma is key to the proof of the theorem.

Lemma 15.2 *If T is ϕ-av, for some $\phi \in (0, 1)$, then $A^T(I - T)A$ is $\frac{1}{2\phi L}$-ism.*

Proof: We have

$$\langle A^T(I - T)Ax - A^T(I - T)Ay, x - y\rangle = \langle (I - T)Ax - (I - T)Ay, Ax - Ay\rangle.$$

Since $I - T$ is $\frac{1}{2\phi}$-ism, we have

$$\langle (I - T)Ax - (I - T)Ay, Ax - Ay\rangle \geq \frac{1}{2\phi}\|(I - T)Ax - (I - T)Ay\|^2.$$

From

$$\|A^T(I - T)Ax - A^T(I - T)Ay\|^2 \leq L\|(I - T)Ax - (I - T)Ay\|^2,$$

it follows that

$$\langle (I - T)Ax - (I - T)Ay, Ax - Ay\rangle \geq \frac{1}{2\phi L}\|A^T(I - T)Ax - A^T(I - T)Ay\|^2.$$

∎

Proof of the Theorem: Assume that z is a solution of the SVIP. The operator $\gamma A^T(I - T)A$ is $\frac{1}{2\gamma\phi L}$-ism. The operator V will be averaged if $\gamma\phi L < 1$, or

$$\gamma < \frac{1}{\phi L} = \frac{2}{(\delta + 1)L}. \tag{15.17}$$

If $\gamma \leq \frac{1}{L}$, then the inequality (15.17) holds for all choices of $\lambda < 2\alpha$.

In similar iterative algorithms, such as the CQ algorithm and the Landweber algorithm (see [58, 59]), the upper bound on γ is $\frac{2}{L}$. We can allow γ to approach $\frac{2}{L}$ here, but only by making δ approach zero, that is, only by taking λ near zero.

Since U is also averaged, the operator S is averaged. Since the intersection of Fix(U) and Fix(V) is not empty, this intersection equals Fix(S). By the Krasnosel'skii-Mann-Opial Theorem 7.1, the iteration $x^{k+1} = Sx^k$ converges to a fixed point x^* of S, which is then a fixed point of both U and V. From $V(x^*) = x^*$ it follows that $A^T(T - I)Ax^* = 0$. We show that $(T - I)Ax^* = 0$. We know that $T(Ax^*) = Ax^* + w$, where $A^Tw = 0$. Also $T(Az) = Az$, since z solves the SVIP. Therefore, we have

$$\|T(Ax^*) - T(Az)\|^2 = \|Ax^* - Az\|^2 + \|w\|^2;$$

but T is non-expansive, so $w = 0$ and $(T - I)Ax^* = 0$. ∎

15.7 Saddle Points

As the title of [151] indicates, the main topic of the paper is saddle points. The main theorem is about convergence of an iterative method for finding saddle points. The saddle-point problem can be turned into a case of the variational inequality problem, which is why the paper contains the theorem on convergence of an iterative algorithm for the VIP that we have already discussed.

The increased complexity of Korpelevich's algorithm is not needed if our goal is to minimize a convex function $f(x)$ over a closed convex set C, when ∇f is L-Lipschitz. In that case, the operator $\frac{1}{L}\nabla f$ is ne, from which it can be shown that it must be fne. Then we can use the averaged operator $T = P_C(I - \gamma\nabla f)$, for $0 < \gamma < \frac{2}{L}$. Similarly, we don't need Korpelevich to solve $z = Nz$, for non-expansive N; we can use the averaged operator $T = (1 - \alpha)I + \alpha N$. However, for saddle-point problems, Korpelevich's method is useful.

15.7.1 Notation and Basic Facts

Let $C \subseteq \mathbb{R}^J$ and $Q \subseteq \mathbb{R}^I$ be closed convex sets. Say that $u^* = (x^*, y^*) \in U$ is a *saddle-point* for the function $f(x, y) : U = C \times Q \to \mathbb{R}$ if, for all $u = (x, y) \in U$, we have

$$f(x^*, y) \leq f(x^*, y^*) \leq f(x, y^*). \tag{15.18}$$

We make the usual assumptions that $f(x, y)$ is convex in x and concave in y, and that the partial derivatives $f_x(x, y)$ and $f_y(x, y)$ are L-Lipschitz. Denote by U^* the set of all saddle points u^*.

It can be shown that $u^* = (x^*, y^*)$ is a saddle point for $f(x, y)$ if and only if

$$\langle f_x(x^*, y^*), x - x^* \rangle \geq 0,$$

for all $x \in C$, and

$$\langle f_y(x^*, y^*), y - y^* \rangle \leq 0,$$

for all $y \in Q$.

15.7.2 The Saddle-Point Problem as a VIP

Define $T : U \to \mathbb{R}^J \times \mathbb{R}^I$ by

$$Tu = (f_x(x, y), -f_y(x, y)).$$

Then $u^* \in U^*$ if and only if

$$\langle Tu^*, u - u^* \rangle \geq 0,$$

for all $u \in U$. The operator T is monotone and L-Lipschitz. Therefore, we can find saddle points by applying the Korpelevich method for finding solutions of the VIP.

Note that if T is a gradient operator, then we must have

$$f(x, y) = h(x) - g(y),$$

where h and g are convex functions. Then (x^*, y^*) is a saddle point if and only if x^* minimizes $h(x)$ over $x \in C$ and y^* minimizes $g(y)$ over $y \in Q$. In this case, the saddle point can be found by solving two independent minimization problems, and Korpelevich's algorithm is not needed.

15.7.3 Example: Convex Programming

In convex programming (CP) we want to minimize a convex function $f : \mathbb{R}^J \to \mathbb{R}$ over all $x \geq 0$ such that $g(x) \leq 0$, where g is also convex. The Lagrangian function is

$$L(x, y) = f(x) + \langle y, g(x) \rangle. \tag{15.19}$$

When the problem is super-consistent, we know that x^* is a solution of CP if and only if there is $y^* \geq 0$ such that

$$L(x^*, y) \leq L(x^*, y^*) \leq L(x, y^*). \tag{15.20}$$

15.7.4 Example: Linear Programming

The primary problem of linear programming, in canonical form, denoted PC, is to minimize $z = c^T x$, subject to $x \geq 0$ and $A^T x \geq b$. The Lagrangian is now

$$L(x, y) = c^T x + y^T(b - A^T x) = c^T x + b^T y - y^T A^T x. \tag{15.21}$$

Therefore, x^* is a solution if and only if there is $y^* \geq 0$ such that (15.20) holds for $L(x, y)$ given by Equation (15.21).

15.7.5 Example: Game Theory

In two-person zero-sum matrix games, the entries A_{mn} of the matrix A are the payoffs from Player Two (P2) to Player One (P1) when P1 plays strategy m and P2 plays strategy n. Optimal randomized strategies p^* and q^* for players P1 and P2, respectively, are probability vectors that satisfy the saddle-point condition

$$f(q^*, p) \leq f(q^*, p^*) \leq f(q, p^*), \tag{15.22}$$

where $f(q, p) = p^T Aq$ for probability vectors p and q.

We could attempt to find the optimal randomized strategies p^* and q^* using Korpelevich's saddle point method; however, the constraint that the p and q be probability vectors may be difficult to implement in the iterative algorithm. There is another way.

A standard approach to showing that optimal randomized strategies exist is to convert the problem into a linear programming problem. Specifically, we first modify A so that all the entries are nonnegative. Then we take b and c to have all entries equal to one. We then minimize $z = c^T x$ over all $x \geq 0$ with $A^T x \geq b$. Because the entries of A are nonnegative, both the primary and dual linear programming problems have feasible solutions, and therefore have optimal solutions, which we denote by x^* and y^*. It follows that $\mu = c^T x^* = b^T y^*$, so that $p^* = \frac{1}{\mu} x^*$ and $q^* = \frac{1}{\mu} y^*$ are probabilities. They are then the optimal randomized strategies.

From this, we see that we can reformulate the search for the game-theory saddle point as a search for a saddle point of the Lagrangian for the linear programming problem. Now the constraints are only that $x \geq 0$ and $y \geq 0$.

15.8 Exercises

Ex. 15.1 *Let* $T : \mathbb{R}^2 \rightarrow \mathbb{R}^2$ *be the operator defined by* $T(x, y) = (-y, x)$. *Show that* T *is a monotone operator, but is not a gradient operator.*

Chapter 16

Set-Valued Functions in Optimization

16.1 Overview

Set-valued mappings play an important role in a number of optimization problems. We examine several of those problems in this chapter. We discuss iterative algorithms for solving these problems and prove convergence.

16.2 Notation and Definitions

Let $f(x) = \iota_C(x)$, the *indicator function* of the set C, which is $+\infty$ for x not in C and zero for x in C. Then

$$\partial\iota_C(x) = N_C(x). \tag{16.1}$$

Most of the time, but not always, we have

$$\partial(f + g)(x) = \partial f(x) + \partial g(x).$$

Consequently, most of the time, but not always, we have

$$N_{A\cap B}(x) = N_A(x) \cap N_B(x); \tag{16.2}$$

see Exercise (16.1). In order for Equation (16.2) to hold, some additional conditions are needed; for example, it is enough to know that the set $A \cap B$ has a nonempty interior (see [27], p. 56, Exercise 10).

The mapping that takes each x to $\partial f(x)$ is a *set-valued function*, or *multi-valued function*. The role that set-valued functions play in optimization is the subject of this chapter. It is common to use the notation $2^{\mathbb{R}^J}$ to denote the collection of all subsets of $\mathbb{R}^J$.

16.3 Basic Facts

If x^* minimizes the function $f(x)$ over all x in $\mathbb{R}^J$, then $0 \in \partial f(x^*)$; if f is differentiable, then $\nabla f(x^*) = 0$. The vector x^* minimizes $f(x)$ over x in C if and only if x^* minimizes the function $f(x) + \iota_C(x)$ over all x in $\mathbb{R}^J$, and so if and only if $0 \in \partial f(x^*) + N_C(x^*)$, which is equivalent to

$$\langle u, c - x^* \rangle \geq 0, \tag{16.3}$$

for all u in $\partial f(x^*)$ and all c in C. If f is differentiable at x^*, then this becomes

$$\langle \nabla f(x^*), c - x^* \rangle \geq 0, \tag{16.4}$$

for all c in C.

Similarly, for each fixed x, y minimizes the function

$$f(t) + \frac{1}{2}\|x - t\|_2^2$$

if and only if

$$0 \in y - x + \partial f(y),$$

or

$$x \in y + \partial f(y).$$

Then we write $y = \text{prox}_f x$. If C is a nonempty closed convex subset of $\mathbb{R}^J$ and $f(x) = \iota_C(x)$, then $\text{prox}_f(x) = P_C x$.

16.4 Monotone Set-Valued Functions

A set-valued function $B : \mathbb{R}^J \to 2^{\mathbb{R}^J}$ is monotone if, for every x and y, and every $u \in B(x)$ and $v \in B(y)$ we have

$$\langle u - v, x - y \rangle \geq 0.$$

A monotone (possibly set-valued) function B is a *maximal monotone* operator if the domain of B cannot be enlarged without the loss of the monotone property.

Let $f : \mathbb{R}^J \to \mathbb{R}$ be convex and differentiable. Then the operator $T = \nabla f$ is monotone. If $f(x)$ is convex, but not differentiable, then $B(x) = \partial f(x)$ is a monotone set-valued function. If A is a non-zero, skew-symmetric matrix, then $Tx = Ax$ is a monotone operator, but is not a gradient operator.

16.5 Resolvents

Let $B : \mathbb{R}^J \to 2^{\mathbb{R}^J}$ be a set-valued mapping. If B is monotone, then $x \in z + B(z)$ and $x \in y + B(y)$ implies that $z = y$, since then $x - z \in B(z)$ and $x - y \in B(y)$, so that

$$0 \leq \langle (x - z) - (x - y), z - y \rangle = -\|z - y\|_2^2.$$

Consequently, the *resolvent operator* for B, defined by

$$J_B = (I + B)^{-1}$$

is single-valued, where

$$J_B(x) = z$$

means that

$$x \in z + B(z).$$

If $B(z) = N_C(z)$ for all z, then $z = J_B(x)$ if and only if $z = P_C(x)$, the orthogonal projection of x onto C; so we have

$$J_{\partial\iota_C} = J_{N_C} = P_C = \text{prox}_{\iota_C}.$$

We know that $z = \text{prox}_f x$ if and only if $x - z \in \partial f(z)$, and so if and only if $x \in (I + \partial f)z$ or, equivalently, $z = J_{\partial f} x$. Therefore,

$$J_{\partial f} = \text{prox}_f.$$

As we shall see shortly, this means that prox_f is fne.

The following theorem is helpful in proving convergence of iterative fixed-point algorithms [99, 57, 100].

Theorem 16.1 *An operator $T : \mathbb{R}^J \to \mathbb{R}^J$ is firmly nonexpansive if and only if $T = J_B$ for some (possibly set-valued) maximal monotone function B.*

We sketch the proof here. Showing that J_B is fne when B is monotone is not difficult. To go the other way, we suppose that T is fne and define $B(x) = T^{-1}\{x\} - x$, where $y \in T^{-1}\{x\}$ means $Ty = x$. Then $J_B = T$. That this function B is monotone follows fairly easily from the fact that $T = J_B$ is fne.

16.6 Split Monotone Variational Inclusion

Let $B_1 : \mathbb{R}^J \to 2^{\mathbb{R}^J}$ and $B_2 : \mathbb{R}^I \to 2^{\mathbb{R}^I}$ be set-valued mappings, $A : \mathbb{R}^J \to \mathbb{R}^I$ a real matrix, and $f : \mathbb{R}^J \to \mathbb{R}^J$ and $g : \mathbb{R}^I \to \mathbb{R}^I$ single-valued operators. Following Moudafi [172], we can pose the *split monotone variational inclusion* problem (SMVIP).

The SMVIP is to find x^* in $\mathbb{R}^J$ such that

$$0 \in f(x^*) + B_1(x^*), \tag{16.5}$$

and

$$0 \in g(Ax^*) + B_2(Ax^*). \tag{16.6}$$

Let C be a closed, nonempty, convex set in $\mathbb{R}^J$. The *normal cone* to C at z is defined to be the empty set if z is not in C, and, if $z \in C$, to be the set $N_C(z)$ given by

$$N_C(z) = \{u | \langle u, c - z\rangle \leq 0, \text{for all}\, c \in C\}. \tag{16.7}$$

Suppose that $C \subseteq \mathbb{R}^J$ and $Q \subseteq \mathbb{R}^I$ are closed nonempty convex sets. If we let $B_1 = N_C$ and $B_2 = N_Q$, then the SMVIP becomes the *split variational inequality* problem (SVIP): find x^* in C such that

$$\langle f(x^*), c - x^*\rangle \geq 0, \tag{16.8}$$

for all $c \in C$, and

$$\langle g(Ax^*), q - Ax^*\rangle \geq 0, \tag{16.9}$$

for all $q \in Q$.

16.7 Solving the SMVIP

We can solve the SMVIP in a way similar to that used to solve the SVIP, by modifying the CGR algorithm. Now we define

$$S = U(I - \gamma A^T(T - I)A),$$

where

$$U = J_{\lambda B_1}(I - \lambda f),$$

and

$$T = J_{\lambda B_2}(I - \lambda g),$$

for $\lambda > 0$. It is easy to show that x^* satisfies Equation (16.5) if and only if x^* is a fixed point of U and Ax^* satisfies Equation (16.6) if and only if Ax^* is a fixed point of T. We have assumed that there is a z that solves the SMVIP, so it follows that z is a fixed point of both U and V, where V is given by

$$V = (I + \gamma A^T(T - I)A).$$

Under the assumption that both B_1 and B_2 are maximal monotone set-valued mappings, we can conclude that both $J_{\lambda B_1}$ and $J_{\lambda B_2}$ are fne operators, and so are av operators. It follows that both U and V are averaged, as well, so that S is averaged.

Now we can argue just as we did in the proof of convergence of the algorithm for the SVIP that the sequence $\{S^k x^0\}$ converges to a fixed point of S, which is then a solution of the SMVIP.

16.8 Special Cases of the SMVIP

There are several problems that can be formulated and solved as special cases of the SMVIP. One example is the *split minimization problem.*

16.8.1 The Split Minimization Problem

Let $f : \mathbb{R}^J \to \mathbb{R}$ and $g : \mathbb{R}^I \to \mathbb{R}$ be lower semicontinuous, convex functions, and C and Q nonempty, closed, convex subsets of $\mathbb{R}^J$ and $\mathbb{R}^I$, respectively. The split minimization problem is to find $x = x^* \in C$ that minimizes $f(x)$ over all $x \in C$, and such that $q = Ax^* \in Q$ minimizes $g(q)$ over all $q \in Q$.

16.9 The Split Common Null-Point Problem

The *split common null-point problem* (SCNPP) [69] is related to the SMVIP. Let $\mathcal{H}_1$ and $\mathcal{H}_2$ be real Hilbert spaces. Let $B_i : \mathcal{H}_1 \to 2^{\mathcal{H}_1}$, for $i = 1, ..., p$, and $F_j : \mathcal{H}_2 \to 2^{\mathcal{H}_2}$, for $j = 1, ..., r$, be set-valued mappings, and $A_j : \mathcal{H}_1 \to \mathcal{H}_2$ be bounded linear operators. The SCNPP is the following: find a point x^* in $\mathcal{H}_1$ such that

$$0 \in \cap_{i=1}^{p} B_i(x^*),$$

and such that, for $y_j^* = A_j(x^*)$, we have

$$0 \in \cap_{j=1}^{r} F_j(y_j^*).$$

16.10 Exercises

Ex. 16.1 *In $\mathbb{R}^2$, let A and B be the closed circles with radius one centered at $(-1, 0)$ and $(1, 0)$, respectively. Show that $N_{A\cap B}((0, 0)) = \mathbb{R}^2$, while $N_A((0, 0)) + N_B((0, 0))$ is the x-axis.*

Chapter 17

Fenchel Duality

17.1 The Legendre-Fenchel Transformation

The duality between convex functions on $\mathbb{R}^J$ and their tangent hyperplanes is made explicit through the Legendre-Fenchel transformation. In this chapter we discuss this transformation, state and prove Fenchel's Duality Theorem, and investigate some of its applications.

Throughout this section $f : C \subseteq \mathbb{R}^J \to \mathbb{R}$ is a closed, proper, convex function defined on a nonempty, closed convex set C.

17.1.1 The Fenchel Conjugate

We say that a function $h(x) : \mathbb{R}^J \to \mathbb{R}$ is *affine* if it has the form $h(x) = \langle a, x\rangle - \gamma$, for some vector a and scalar γ. If $\gamma = 0$, then we call the function *linear*. A function such as $f(x) = 5x + 2$ is commonly called a linear function in algebra classes, but, according to our definitions, it should be called an affine function.

For each fixed vector a in $\mathbb{R}^J$, the affine function $h(x) = \langle a, x \rangle - \gamma$ is beneath the function $f(x)$ if $f(x) - h(x) \geq 0$, for all x; that is,

$$f(x) - \langle a, x \rangle + \gamma \geq 0,$$

or

$$\gamma \geq \langle a, x \rangle - f(x). \tag{17.1}$$

This leads us to the following definition, involving the maximum of the right side of the inequality in (17.1), for each fixed a.

Definition 17.1 *The* conjugate function *associated with f is the function*

$$f^*(a) = \sup_{x \in C}(\langle a, x \rangle - f(x)). \tag{17.2}$$

We then define C^* to be the set of all a for which $f^*(a)$ is finite. For each fixed a, the value $f^*(a)$ is the smallest value of γ for which the affine function $h(x) = \langle a, x \rangle - \gamma$ is beneath $f(x)$ for $x \in C$. The passage from f to f^* is the *Legendre-Fenchel Transformation.*

For example, suppose that $f(x) = \frac{1}{2}x^2$. The function $h(x) = ax + b$ is beneath $f(x)$ for all x if

$$ax + b \leq \frac{1}{2}x^2,$$

for all x. Equivalently,

$$b \leq \frac{1}{2}x^2 - ax,$$

for all x. Then b must not exceed the minimum of the right side, which is $-\frac{1}{2}a^2$ and occurs when $x - a = 0$, or $x = a$. Therefore, we have

$$\gamma = -b \geq \frac{1}{2}a^2.$$

The smallest value of γ for which this is true is $\gamma = \frac{1}{2}a^2$, so we have $f^*(a) = \frac{1}{2}a^2$.

17.1.2 The Conjugate of the Conjugate

Now we repeat this process with $f^*(a)$ in the role of $f(x)$. For each fixed vector x, the affine function $c(a) = \langle a, x \rangle - \gamma$ is beneath the function $f^*(a)$ if $f^*(a) - c(a) \geq 0$, for all $a \in C^*$; that is,

$$f^*(a) - \langle a, x \rangle + \gamma \geq 0,$$

or

$$\gamma \geq \langle a, x \rangle - f^*(a). \tag{17.3}$$

This leads us to the following definition, involving the maximum of the right side of the inequality in (17.3), for each fixed x.

Definition 17.2 *The conjugate function associated with f^* is the function*

$$f^{**}(x) = \sup_a(\langle a, x\rangle - f^*(a)). \tag{17.4}$$

For each fixed x, the value $f^{**}(x)$ is the smallest value of γ for which the affine function $c(a) = \langle a, x\rangle - \gamma$ is beneath $f^*(a)$.

Applying the Separation Theorem to the epigraph of the closed, proper, convex function $f(x)$, it can be shown ([192], Theorem 12.1) that $f(x)$ is the point-wise supremum of all the affine functions beneath $f(x)$; that is,

$$f(x) = \sup_{a,\gamma}\{h(x)|f(x) \geq h(x)\}.$$

Therefore,

$$f(x) = \sup_a \Big(\langle a, x\rangle - f^*(a)\Big).$$

This says that

$$f^{**}(x) = f(x). \tag{17.5}$$

If $f(x)$ is a differentiable function, then, for each fixed a, the function

$$g(x) = \langle a, x\rangle - f(x)$$

attains its minimum when

$$0 = \nabla g(x) = a - \nabla f(x),$$

which says that $a = \nabla f(x)$.

17.1.3 Some Examples of Conjugate Functions

- The exponential function $f(x) = \exp(x) = e^x$ has conjugate

  $$\exp^*(a) = a \log a - a, \tag{17.6}$$

  if $a > 0$, 0 if $a = 0$, and $+\infty$ if $a < 0$.

- The function $f(x) = -\log x$, for $x > 0$, has the conjugate function $f^*(a) = -1 - \log(-a)$, for $a < 0$.

- The function $f(x) = \frac{|x|^p}{p}$ has conjugate $f^*(a) = \frac{|a|^q}{q}$, where $p > 0$, $q > 0$, and $\frac{1}{p} + \frac{1}{q} = 1$. Therefore, the function $f(x) = \frac{1}{2}\|x\|^2$ is its own conjugate, that is, $f^*(a) = \frac{1}{2}\|a\|^2$.

- Let A be a real symmetric positive-definite matrix and

  $$f(x) = \frac{1}{2}\langle Ax, x\rangle.$$

  Then

  $$f^*(a) = \frac{1}{2}\langle A^{-1}a, a\rangle.$$

- Let $i_C(x)$ be the *indicator function* of the closed convex set C, that is, $i_C(x) = 0$, if $x \in C$, and ∞ otherwise.Then

$$i_C^*(a) = \sup_{x \in C} \langle a, x \rangle,$$

which is the *support function* of the set C, usually denoted $\sigma_C(a)$.

- Let $C \subseteq \mathbb{R}^J$ be nonempty, closed and convex. The *gauge function* of C is

$$\gamma_C(x) = \inf\{\lambda \geq 0 \,|x \in \lambda C\}.$$

If $C = B$, the unit ball of $\mathbb{R}^J$, then $\gamma_B(x) = \|x\|_2$. For each C define the *polar set* for C by

$$C^0 = \{z|\langle z, c\rangle \leq 1, \text{for all}\, c \in C\}.$$

Then

$$\gamma_C^* = \iota_{C^0}.$$

- Let $C = \{x| \; ||x||_2 \leq 1\}$, so that the function $\phi(a) = \|a\|_2$ satisfies

$$\phi(a) = \sup_{x \in C} \langle a, x \rangle.$$

Then

$$\phi(a) = \sigma_C(a) = i_C^*(a).$$

Therefore,

$$\phi^*(x) = \sigma_C^*(x) = i_C^{**}(x) = i_C(x).$$

17.1.4 Infimal Convolution Again

The infimal convolution and deconvolution are related to the Fenchel conjugate; specifically, under suitable conditions, we have

$$f \oplus g = (f^* + g^*)^*,$$

and

$$f \ominus g = (f^* - g^*)^*.$$

See Lucet [161] for details.

17.1.5 Conjugates and Sub-gradients

We know from the definition of $f^*(a)$ that

$$f^*(a) \geq \langle a, z\rangle - f(z),$$

for all z, and, moreover, $f^*(a)$ is the supremum of these values, taken over all z. If a is a member of the sub-differential $\partial f(x)$, then, for all z, we have

$$f(z) \geq f(x) + \langle a, z - x\rangle,$$

so that

$$\langle a, x\rangle - f(x) \geq \langle a, z\rangle - f(z).$$

It follows that

$$f^*(a) = \langle a, x\rangle - f(x),$$

so that

$$f(x) + f^*(a) = \langle a, x\rangle.$$

If $f(x)$ is a differentiable convex function, then a is in the sub-differential $\partial f(x)$ if and only if $a = \nabla f(x)$. Then we can say

$$f(x) + f^*(\nabla f(x)) = \langle \nabla f(x), x\rangle. \tag{17.7}$$

If $a = \nabla f(x_1)$ and $a = \nabla f(x_2)$, then the function

$$g(x) = \langle a, x\rangle - f(x)$$

attains its minimum value at $x = x_1$ and at $x = x_2$, so that

$$f^*(a) = \langle a, x_1\rangle - \nabla f(x_1) = \langle a, x_2\rangle - f(x_2).$$

Let us denote by $x = (\nabla f)^{-1}(a)$ any x for which $\nabla f(x) = a$. Then the conjugate of the differentiable function $f : C \subseteq \mathbb{R}^J \to \mathbb{R}$ can then be defined as follows [192]. Let D be the image of the set C under the mapping ∇f. Then, for all $a \in D$, define

$$f^*(a) = \langle a, (\nabla f)^{-1}(a)\rangle - f((\nabla f)^{-1}(a)). \tag{17.8}$$

The formula in Equation (17.8) is also called the Legendre Transform.

17.1.6 The Conjugate of a Concave Function

A function $g : D \subseteq \mathbb{R}^J \to \mathbb{R}$ is *concave* if $f(x) = -g(x)$ is convex. One might think that the conjugate of a concave function g is simply the negative of the conjugate of $-g$, but not quite.

The affine function $h(x) = \langle a, x\rangle - \gamma$ is above the concave function $g(x)$ if $h(x) - g(x) \geq 0$, for all $x \in D$; that is,

$$\langle a, x\rangle - \gamma - g(x) \geq 0,$$

or

$$\gamma \leq \langle a, x\rangle - g(x). \tag{17.9}$$

The conjugate function associated with g is the function

$$g^*(a) = \inf_x(\langle a, x\rangle - g(x)). \tag{17.10}$$

For each fixed a, the value $g^*(a)$ is the largest value of γ for which the affine function $h(x) = \langle a, x\rangle - \gamma$ is above $g(x)$.

It follows, using $f(x) = -g(x)$, that

$$g^*(a) = \inf_x(\langle a, x\rangle + f(x)) = -\sup_x(\langle -a, x\rangle - f(x)) = -f^*(-a).$$

17.2 Fenchel's Duality Theorem

Let $f(x)$ be a proper convex function on $C \subseteq \mathbb{R}^J$ and $g(x)$ a proper concave function on $D \subseteq \mathbb{R}^J$, where C and D are closed convex sets with nonempty intersection. Fenchel's Duality Theorem deals with the problem of minimizing the difference $f(x) - g(x)$ over $x \in C \cap D$.

We know from our discussion of conjugate functions and differentiability that

$$-f^*(a) \leq f(x) - \langle a, x\rangle,$$

and

$$g^*(a) \leq \langle a, x\rangle - g(x).$$

Therefore,

$$f(x) - g(x) \geq g^*(a) - f^*(a),$$

for all x and a, and so

$$\inf_x \Big(f(x) - g(x)\Big) \geq \sup_a \Big(g^*(a) - f^*(a)\Big).$$

We let C^* be the set of all a such that $f^*(a)$ is finite, with D^* similarly defined.

The Fenchel Duality Theorem, in its general form, as found in [162] and [192], is as follows.

Theorem 17.1 *Assume that $C \cap D$ has points in the relative interior of both C and D, and that either the epigraph of f or that of g has nonempty interior. Suppose that*

$$\mu = \inf_{x \in C \cap D} \Big(f(x) - g(x) \Big)$$

is finite. Then

$$\mu = \inf_{x \in C \cap D} \Big(f(x) - g(x) \Big) = \max_{a \in C^* \cap D^*} \Big(g^*(a) - f^*(a) \Big),$$

where the maximum on the right is achieved at some $a_0 \in C^* \cap D^*$.

If the infimum on the left is achieved at some $x_0 \in C \cap D$, then

$$\max_{x \in C} \Big(\langle x, a_0 \rangle - f(x) \Big) = \langle x_0, a_0 \rangle - f(x_0),$$

and

$$\min_{x \in D} \Big(\langle x, a_0 \rangle - g(x) \Big) = \langle x_0, a_0 \rangle - g(x_0).$$

The conditions on the interiors are needed to make use of subdifferentials. For simplicity, we shall limit our discussion to the case of differentiable $f(x)$ and $g(x)$.

17.2.1 Fenchel's Duality Theorem: Differentiable Case

We suppose now that there is $x_0 \in C \cap D$ such that

$$\inf_{x \in C \cap D} (f(x) - g(x)) = f(x_0) - g(x_0),$$

and that

$$\nabla(f - g)(x_0) = 0,$$

or

$$\nabla f(x_0) = \nabla g(x_0). \tag{17.11}$$

Let $\nabla f(x_0) = a_0$. From the equation

$$f(x) + f^*(\nabla f(x)) = \langle \nabla f(x), x \rangle$$

and Equation (17.11),we have

$$f(x_0) - g(x_0) = g^*(a_0) - f^*(a_0),$$

from which it follows that

$$\inf_{x \in C \cap D} (f(x) - g(x)) = \sup_{a \in C^* \cap D^*} (g^*(a) - f^*(a)).$$

This is Fenchel's Duality Theorem.

17.2.2 Optimization over Convex Subsets

Suppose now that $f(x)$ is convex and differentiable on $\mathbb{R}^J$, but we are only interested in its values on the nonempty closed convex set C. Then we redefine $f(x) = +\infty$ for x not in C. The affine function $h(x) = \langle a, x\rangle - \gamma$ is beneath $f(x)$ for all x if and only if it is beneath $f(x)$ for $x \in C$. This motivates our defining the conjugate function now as

$$f^*(a) = \sup_{x\in C}\langle a, x\rangle - f(x).$$

Similarly, let $g(x)$ be concave on D and $g(x) = -\infty$ for x not in D. Then we define

$$g^*(a) = \inf_{x\in D}\langle a, x\rangle - g(x).$$

Let

$$C^* = \{a|\, f^*(a) < +\infty\},$$

and define D^* similarly. We can use Fenchel's Duality Theorem to minimize the difference $f(x) - g(x)$ over the intersection $C \cap D$.

To illustrate the use of Fenchel's Duality Theorem, consider the problem of minimizing the convex function $f(x)$ over the convex set D. Let $C = \mathbb{R}^J$ and $g(x) = 0$, for all x. Then

$$f^*(a) = \sup_{x\in C}\Big(\langle a, x\rangle - f(x)\Big) = \sup_{x}\Big(\langle a, x\rangle - f(x)\Big),$$

and

$$g^*(a) = \inf_{x\in D}\Big(\langle a, x\rangle - g(x)\Big) = \inf_{x\in D}\langle a, x\rangle.$$

The supremum is unconstrained and the infimum is with respect to a linear functional. Then, by Fenchel's Duality Theorem, we have

$$\max_{a\in C^*\cap D^*}(g^*(a) - f^*(a)) = \inf_{x\in D} f(x).$$

17.3 An Application to Game Theory

In this section we apply the Fenchel Duality Theorem to prove the Min-Max Theorem for two-person zero-sum matrix games.

17.3.1 Pure and Randomized Strategies

In a two-person game, the first player selects a row of the matrix A, say i, and the second player selects a column of A, say j. The second player pays the first player A_{ij}. If some $A_{ij} < 0$, then this means that the first player pays the second. As we discussed previously, there need not be optimal pure strategies for the two players and it may be sensible for them, over the long run, to select their strategies according to some random mechanism. The issues then are which vectors of probabilities will prove optimal and do such optimal probability vectors always exist. The Min-Max Theorem, also known as the Fundamental Theorem of Game Theory, asserts that such optimal probability vectors always exist.

17.3.2 The Min-Max Theorem

In [162], Luenberger uses the Fenchel Duality Theorem to prove the Min-Max Theorem for two-person games. His formulation is in Banach spaces, while we shall limit our discussion to finite-dimensional Euclidean spaces.

Let A be an I by J pay-off matrix, whose entries represent the payoffs from the second player to the first. Let

$$P = \{p = (p_1, ..., p_I) \,|\, p_i \geq 0, \sum_{i=1}^{I} p_i = 1\},$$

$$S = \{s = (s_1, ..., s_I) \,|\, s_i \geq 0, \sum_{i=1}^{I} s_i \leq 1\},$$

and

$$Q = \{q = (q_1, ..., q_J) \,|\, q_j \geq 0, \sum_{j=1}^{J} q_j = 1\}.$$

The set S is the convex hull of the set P.

The first player selects a vector p in P and the second selects a vector q in Q. The expected pay-off to the first player is

$$E = \langle p, Aq \rangle.$$

Let

$$m_0 = \max_{p \in P} \min_{q \in Q} \langle p, Aq \rangle,$$

and

$$m^0 = \min_{q \in Q} \max_{p \in P} \langle p, Aq \rangle.$$

Clearly, we have

$$\min_{q\in Q}\langle p, Aq\rangle \le \langle p, Aq\rangle \le \max_{p\in P}\langle p, Aq\rangle,$$

for all $p \in P$ and $q \in Q$. It follows that $m_0 \le m^0$. We show that $m_0 = m^0$.

Define

$$f(x) = \max_{p\in P}\langle p, x\rangle,$$

which is equivalent to

$$f(x) = \max_{s\in S}\langle s, x\rangle.$$

Then f is convex and continuous on $\mathbb{R}^I$. We want $\min_{q\in Q} f(Aq)$.

We apply Fenchel's Duality Theorem, with $f = f$, $g = 0$, $D = A(Q)$, and $C = \mathbb{R}^I$. Now we have

$$\inf_{x\in C\cap D}(f(x) - g(x)) = \min_{q\in Q} f(Aq).$$

We claim that the following are true:

1. $D^* = \mathbb{R}^I$;

2. $g^*(a) = \min_{q\in Q}\langle a, Aq\rangle$;

3. $C^* = S$;

4. $f^*(a) = 0$, for all a in S.

The first two claims are immediate. To prove the third one, we take a vector $a \in \mathbb{R}^I$ that is not in S. Then, by the separation theorem, we can find $x \in \mathbb{R}^I$ and $\alpha > 0$ such that

$$\langle x, a\rangle > \alpha + \langle x, s\rangle,$$

for all $s \in S$. Then

$$\langle x, a\rangle - \max_{s\in S}\langle x, s\rangle \ge \alpha > 0.$$

Now take $k > 0$ large and $y = kx$. Since

$$\langle y, s\rangle = k\langle x, s\rangle,$$

we know that

$$\langle y, a\rangle - \max_{s\in S}\langle y, s\rangle = \langle y, a\rangle - f(y) > 0$$

and can be made arbitrarily large by taking $k > 0$ large. It follows that $f^*(a)$ is not finite if a is not in S, so that $C^* = S$.

As for the fourth claim, if $a \in S$, then

$$\langle y, a\rangle - \max_{s\in S}\langle y, s\rangle$$

achieves its maximum value of zero at $y = 0$, so $f^*(a) = 0$.

Finally, we have

$$\min_{q\in Q}\max_{p\in P}\langle p, Aq\rangle = \min_{q\in Q} f(Aq) = \max_{a\in S} g^*(a) = \max_{a\in S}\min_{q\in Q}\langle p, Aq\rangle.$$

Therefore,

$$\min_{q\in Q}\max_{p\in P}\langle p, Aq\rangle = \max_{p\in P}\min_{q\in Q}\langle p, Aq\rangle.$$

17.4 Exercises

Ex. 17.1 *Show that the exponential function $f(x) = \exp(x) = e^x$ has conjugate*

$$\exp^*(a) = a\log a - a, \tag{17.12}$$

if $a > 0$, 0 if $a = 0$, and $+\infty$ if $a < 0$.

Ex. 17.2 *Show that the function $f(x) = -\log x$, for $x > 0$, has the conjugate function $f^*(a) = -1 - \log(-a)$, for $a < 0$.*

Ex. 17.3 *Show that the function $f(x) = \frac{|x|^p}{p}$ has conjugate $f^*(a) = \frac{|a|^q}{q}$, where $p > 0$, $q > 0$, and $\frac{1}{p} + \frac{1}{q} - 1$. Therefore, the function $f(x) = \frac{1}{2}\|x\|_2^2$ is its own conjugate, that is, $f^*(a) = \frac{1}{2}\|a\|_2^2$.*

Ex. 17.4 *Let A be a real symmetric positive-definite matrix and*

$$f(x) = \frac{1}{2}\langle Ax, x\rangle.$$

Show that

$$f^*(a) = \frac{1}{2}\langle A^{-1}a, a\rangle.$$

Hints: Find $\nabla f(x)$ and use Equation (17.8).

Chapter 18

Compressed Sensing

18.1 Compressed Sensing

One area that has attracted much attention lately is *compressed sensing* or *compressed sampling* (CS) [111]. For applications such as medical imaging, CS may provide a means of reducing radiation dosage to the patient without sacrificing image quality. An important aspect of CS is finding sparse solutions of under-determined systems of linear equations, which can often be accomplished by one-norm minimization. Perhaps the best reference to date on CS is [36].

The objective in CS is exploit sparseness to reconstruct a vector f in $\mathbb{R}^J$ from relatively few linear functional measurements [111].

Let $U = \{u^1, u^2, ..., u^J\}$ and $V = \{v^1, v^2, ..., v^J\}$ be two orthonormal bases for $\mathbb{R}^J$, with all members of $\mathbb{R}^J$ represented as column vectors. For $i = 1, 2, ..., J$, let

$$\mu_i = \max_{1 \le j \le J} \{|\langle u^i, v^j \rangle|\}$$

and

$$\mu(U, V) = \max\{\mu_i \,|i = 1, ..., I\}.$$

We know from Cauchy's Inequality that

$$|\langle u^i, v^j \rangle| \le 1,$$

and from Parseval's Equation

$$\sum_{j=1}^{J} |\langle u^i, v^j \rangle|^2 = ||u^i||_2^2 = 1.$$

Therefore, we have

$$\frac{1}{\sqrt{J}} \leq \mu(U, V) \leq 1.$$

The quantity $\mu(U, V)$ is the *coherence* measure of the two bases; the closer $\mu(U, V)$ is to the lower bound of $\frac{1}{\sqrt{J}}$, the more *incoherent* the two bases are.

Let f be a fixed member of $\mathbb{R}^J$; we expand f in the V basis as

$$f = x_1 v^1 + x_2 v^2 + ... + x_J v^J.$$

We say that the coefficient vector $x = (x_1, ..., x_J)$ is s-sparse if s is the number of non-zero x_j.

If s is small, most of the x_j are zero, but since we do not know which ones these are, we would have to compute all the linear functional values

$$x_j = \langle f, v^j \rangle$$

to recover f exactly. In fact, the smaller s is, the harder it would be to learn anything from randomly selected x_j, since most would be zero. The idea in CS is to obtain measurements of f with members of a different orthonormal basis, which we call the U basis. If the members of U are very much like the members of V, then nothing is gained. But, if the members of U are quite unlike the members of V, then each inner product measurement

$$y_i = \langle f, u^i \rangle = f^T u^i$$

should tell us something about f. If the two bases are sufficiently incoherent, then relatively few y_i values should tell us quite a bit about f. Specifically, we have the following result due to Candès and Romberg [74]: suppose the coefficient vector x for representing f in the V basis is s-sparse. Select uniformly randomly $M \leq J$ members of the U basis and compute the measurements $y_i = \langle f, u^i \rangle$. Then, if M is sufficiently large, it is highly probable that $z = x$ also solves the problem of minimizing the one-norm

$$||z||_1 = |z_1| + |z_2| + ... + |z_J|,$$

subject to the conditions

$$y_i = \langle g, u^i \rangle = g^T u^i,$$

for those M randomly selected u^i, where

$$g = z_1 v^1 + z_2 v^2 + ... + z_J v^J.$$

The smaller $\mu(U, V)$ is, the smaller the M is permitted to be without reducing the probability of perfect reconstruction.

18.2 Sparse Solutions

Suppose that A is a real M by N matrix, with $M < N$, and that the linear system $Ax = b$ has infinitely many solutions. For any vector x, we define the *support* of x to be the subset S of $\{1, 2, ..., N\}$ consisting of those n for which the entries $x_n \neq 0$. For any under-determined system $Ax = b$, there will, of course, be at least one solution of minimum support, that is, for which $s = |S|$, the size of the support set S, is minimum. However, finding such a maximally sparse solution requires combinatorial optimization, and is known to be computationally difficult. It is important, therefore, to have a computationally tractable method for finding maximally sparse solutions.

Consider the problem P_0: among all solutions x of the consistent system $b = Ax$, find one, call it $\hat{x}$, that is maximally sparse, that is, has the minimum number of non-zero entries. Obviously, there will be at least one such solution having minimal support, but finding one, however, is a combinatorial optimization problem and is generally NP-hard.

18.3 Minimum One-Norm Solutions

Instead, we can seek a *minimum one-norm* solution x^*, that is, solve the problem P_1: minimize

$$||x||_1 = \sum_{n=1}^{N} |x_n|,$$

subject to $Ax = b$. As we shall see shortly, problem P_1 can be formulated as a linear programming problem, so is more easily solved. The big questions are: when does P_1 have a unique solution x^*, and when is $x^* = \hat{x}$? The problem P_1 will have a unique solution if and only if A is such that the one-norm satisfies

$$||x^*||_1 < ||x^* + v||_1,$$

for all non-zero v in the null space of A.

If the vector x is required to be nonnegative, then the one-norm is simply the sum of the entries, and minimizing the one-norm subject to $Ax = b$ becomes a linear programming problem. This is the situation in applications involving image reconstruction. Generally, though, the vector x cannot be assumed to be nonnegative. The one-norm is not a linear functional of x, but the problem can still be converted into a linear programming problem.

The entries of x need not be nonnegative, so the problem is not yet a linear programming problem. Let

$$B = \begin{bmatrix} A & -A \end{bmatrix},$$

and consider the linear programming problem of minimizing the function

$$c^T z = \sum_{j=1}^{2J} z_j,$$

subject to the constraints $z \geq 0$, and $Bz = b$. Let z^* be the solution. We write

$$z^* = \begin{bmatrix} u^* \\ v^* \end{bmatrix}.$$

Then, as we shall see, $x^* = u^* - v^*$ minimizes the one-norm, subject to $Ax = b$.

First, we show that $u_j^* v_j^* = 0$, for each j. If, say, there is a j such that $0 < v_j^* \leq u_j^*$, then we can create a new vector z from z^* by replacing the old u_j^* with $u_j^* - v_j^*$ and the old v_j^* with zero, while maintaining $Bz = b$. But then, since $u_j^* - v_j^* < u_j^* + v_j^*$, it follows that $c^T z < c^T z^*$, which is a contradiction. Consequently, we have $\|x^*\|_1 = c^T z^*$.

Now we select any x with $Ax = b$. Write $u_j = x_j$, if $x_j \geq 0$, and $u_j = 0$, otherwise. Let $v_j = u_j - x_j$, so that $x = u - v$. Then let

$$z = \begin{bmatrix} u \\ v \end{bmatrix}.$$

Then $b = Ax = Bz$, and $c^T z = \|x\|_1$. Therefore

$$\|x^*\|_1 = c^T z^* \leq c^T z = \|x\|_1,$$

and x^* must be a minimum one-norm solution.

18.3.1 Why the One-Norm?

When a system of linear equations $Ax = b$ is under-determined, we can find the *minimum-two-norm solution* that minimizes the square of the two-norm,

$$||x||_2^2 = \sum_{n=1}^{N} x_n^2,$$

subject to $Ax = b$. One drawback to this approach is that the two-norm penalizes relatively large values of x_n much more than the smaller ones, so tends to provide non-sparse solutions. Alternatively, we may seek the solution x^* for which the one-norm,

$$||x||_1 = \sum_{n=1}^{N} |x_n|,$$

is minimized. The one-norm still penalizes relatively large entries x_n more than the smaller ones, but much less than the two-norm does. As a result, it often happens that the minimum one-norm solution actually solves P_0 as well.

18.3.2 Comparison with the PDFT

The PDFT approach [41, 42] to solving the under-determined system $Ax = b$ is to select weights $w_n > 0$ and then to find the solution $\tilde{x}$ that minimizes the weighted two-norm given by

$$\sum_{n=1}^{N} |x_n|^2 w_n.$$

Our intention is to select weights w_n so that w_n^{-1} is reasonably close to $|x_n^*|$; consider, therefore, what happens when $w_n^{-1} = |x_n^*|$. We claim that $\tilde{x}$ is also a minimum-one-norm solution.

To see why this is true, note that, for any x, we have

$$\sum_{n=1}^{N} |x_n| = \sum_{n=1}^{N} \frac{|x_n|}{\sqrt{|x_n^*|}} \sqrt{|x_n^*|}$$

$$\leq \sqrt{\sum_{n=1}^{N} \frac{|x_n|^2}{|x_n^*|}} \sqrt{\sum_{n=1}^{N} |x_n^*|}.$$

Therefore,

$$\sum_{n=1}^{N} |\tilde{x}_n| \leq \sqrt{\sum_{n=1}^{N} \frac{|\tilde{x}_n|^2}{|x_n^*|}} \sqrt{\sum_{n=1}^{N} |x_n^*|}$$

$$\leq \sqrt{\sum_{n=1}^{N} \frac{|x_n^*|^2}{|x_n^*|}} \sqrt{\sum_{n=1}^{N} |x_n^*|} = \sum_{n=1}^{N} |x_n^*|.$$

Therefore, $\tilde{x}$ also minimizes the one-norm.

18.3.3 Iterative Reweighting

Let x denote the truth. We want each weight w_n to be a good prior estimate of the reciprocal of $|x_n|$. Because we do not yet know x, we may take a sequential-optimization approach, beginning with weights $w_n^0 > 0$, finding the PDFT solution using these weights, then using this PDFT solution to get a (we hope!) better choice for the weights, and so on. This sequential approach was successfully implemented in the early 1980's by Michael Fiddy and his students [127].

In [75], the same approach is taken, but with respect to the one-norm. Since the one-norm still penalizes larger values disproportionately, balance can be achieved by minimizing a weighted-one-norm, with weights close to the reciprocals of the $|x_n^*|$. Again, not yet knowing x^*, they employ a sequential approach, using the previous minimum-weighted-one-norm solution to obtain the new set of weights for the next minimization. At each step of the sequential procedure, the previous reconstruction is used to estimate the true support of the desired solution.

It is interesting to note that an on-going debate among users of the PDFT concerns the nature of the prior weighting. Again, let x be the truth. Does w_n approximate $|x_n|^{-1}$ or $|x_n|^{-2}$? This is close to the issue treated in [75], the use of a weight in the minimum-one-norm approach.

It should be noted again that finding a sparse solution is not usually the goal in the use of the PDFT, but the use of the weights has much the same effect as using the one-norm to find sparse solutions: to the extent that the weights approximate the entries of $|x^*|^{-1}$, their use reduces the penalty associated with the larger entries of an estimated solution.

18.4 Why Sparseness?

One obvious reason for wanting sparse solutions of $Ax = b$ is that we have prior knowledge that the desired solution is sparse. Such a problem arises in signal analysis from Fourier-transform data. In other cases, such as in the reconstruction of locally constant signals, it is not the signal itself, but its discrete derivative, that is sparse.

18.4.1 Signal Analysis

Suppose that our signal $f(t)$ is known to consist of a small number of complex exponentials, so that $f(t)$ has the form

$$f(t) = \sum_{j=1}^{J} a_j e^{i\omega_j t},$$

for some small number of frequencies ω_j in the interval $[0, 2\pi)$. For $n = 0, 1, ..., N-1$, let $f_n = f(n)$, and let f be the N-vector with entries f_n; we assume that J is much smaller than N. The discrete (vector) Fourier transform of f is the vector $\hat{f}$ having the entries

$$\hat{f}_k = \frac{1}{\sqrt{N}} \sum_{n=0}^{N-1} f_n e^{2\pi ikn/N},$$

for $k = 0, 1, ..., N-1$; we write $\hat{f} = Ef$, where E is the N by N matrix with entries $E_{kn} = \frac{1}{\sqrt{N}} e^{2\pi ikn/N}$. If N is large enough, we may safely assume that each of the ω_j is equal to one of the frequencies $2\pi ik$ and that the vector $\hat{f}$ is J-sparse. The question now is: How many values of $f(n)$ do we need to calculate in order to be sure that we can recapture $f(t)$ exactly? We have the following theorem [73]:

Theorem 18.1 *Let N be prime. Let S be any subset of $\{0, 1, ..., N-1\}$ with $|S| \geq 2J$. Then the vector $\hat{f}$ can be uniquely determined from the measurements f_n for n in S.*

We know that

$$f = E^\dagger \hat{f},$$

where $E^\dagger$ is the conjugate transpose of the matrix E. The point here is that, for any matrix R obtained from the identity matrix I by deleting $N - |S|$ rows, we can recover the vector $\hat{f}$ from the measurements Rf.

If N is not prime, then the assertion of the theorem may not hold, since we can have $n = 0 \bmod N$, without $n = 0$. However, the assertion remains valid for most sets of J frequencies and most subsets S of indices; therefore, with high probability, we can recover the vector $\hat{f}$ from Rf.

Note that the matrix E is *unitary*, that is, $E^\dagger E = I$, and, equivalently, the columns of E form an orthonormal basis for $\mathbb{C}^N$. The data vector is

$$b = Rf = RE^\dagger \hat{f}.$$

In this example, the vector f is not sparse, but can be represented sparsely in a particular orthonormal basis, namely as $f = E^\dagger \hat{f}$, using a sparse vector $\hat{f}$ of coefficients. The *representing basis* then consists of the columns of the matrix $E^\dagger$. The measurements pertaining to the vector f are the values f_n, for n in S. Since f_n can be viewed as the inner product of f with δ^n, the nth column of the identity matrix I, that is,

$$f_n = \langle \delta^n, f \rangle,$$

the columns of I provide the so-called *sampling basis.* With $A = RE^\dagger$ and $x = \hat{f}$, we then have

$$Ax = b,$$

with the vector x sparse. It is important for what follows to note that the matrix A is random, in the sense that we choose which rows of I to use to form R.

18.4.2 Locally Constant Signals

Suppose now that the function $f(t)$ is locally constant, consisting of some number of horizontal lines. We discretize the function $f(t)$ to get the vector $f = (f(0), f(1), ..., f(N))^T$. The discrete derivative vector is $g = (g_1, g_2, ..., g_N)^T$, with

$$g_n = f(n) - f(n-1).$$

Since $f(t)$ is locally constant, the vector g is sparse. The data we will have will not typically be values $f(n)$. The goal will be to recover f from M linear functional values pertaining to f, where M is much smaller than N. We shall assume, from now on, that we have measured, or can estimate, the value $f(0)$.

Our M by 1 data vector d consists of measurements pertaining to the vector f:

$$d_m = \sum_{n=0}^{N} H_{mn} f_n,$$

for $m = 1, ..., M$, where the H_{mn} are known. We can then write

$$d_m = f(0)\Big(\sum_{n=0}^{N} H_{mn}\Big) + \sum_{k=1}^{N}\Big(\sum_{j=k}^{N} H_{mj}\Big) g_k.$$

Since $f(0)$ is known, we can write

$$b_m = d_m - f(0)\Big(\sum_{n=0}^{N} H_{mn}\Big) = \sum_{k=1}^{N} A_{mk} g_k,$$

where

$$A_{mk} = \sum_{j=k}^{N} H_{mj}.$$

The problem is then to find a sparse solution of $Ax = g$. As in the previous example, we often have the freedom to select the linear functions, that is, the values H_{mn}, so the matrix A can be viewed as random.

18.4.3 Tomographic Imaging

The reconstruction of tomographic images is an important aspect of medical diagnosis, and one that combines aspects of both of the previous examples. The data one obtains from the scanning process can often be interpreted as values of the Fourier transform of the desired image; this is precisely the case in magnetic-resonance imaging, and approximately true for x-ray transmission tomography, positron-emission tomography (PET) and single-photon emission tomography (SPECT). The images one encounters in medical diagnosis are often approximately locally constant, so the associated array of discrete partial derivatives will be sparse. If this sparse derivative array can be recovered from relatively few Fourier-transform values, then the scanning time can be reduced.

We turn now to the more general problem of compressed sampling.

18.5 Compressed Sampling

Our goal is to recover the vector $f = (f_1, ..., f_N)^T$ from M linear functional values of f, where M is much less than N. In general, this is not possible without prior information about the vector f. In compressed sampling, the prior information concerns the sparseness of either f itself, or another vector linearly related to f.

Let U and V be unitary N by N matrices, so that the column vectors of both U and V form orthonormal bases for $\mathbb{C}^N$. We shall refer to the bases associated with U and V as the *sampling basis* and the *representing basis*, respectively. The first objective is to find a unitary matrix V so that $f = Vx$, where x is sparse. Then we want to find a second unitary matrix U such that, when an M by N matrix R is obtained from U by deleting rows, the sparse vector x can be determined from the data $b = RVx = Ax$. Theorems in compressed sensing describe properties of the matrices U and V such that, when R is obtained from U by a random selection of the rows of U, the vector x will be uniquely determined, with high probability, as the unique solution that minimizes the one-norm.

Appendix A

Bregman-Legendre Functions

In [13] Bauschke and Borwein show convincingly that the Bregman-Legendre functions provide the proper context for the discussion of Bregman projections onto closed convex sets. The summary here follows closely the discussion given in [13].

A.1 Essential Smoothness and Essential Strict Convexity

Following [192] we say that a closed proper convex function f is *essentially smooth* if int$D$ is not empty, f is differentiable on int$D$ and $x^n \in$ int$D$, with $x^n \to x \in$ bd$D$, implies that $||\nabla f(x^n)||_2 \to +\infty$. Here

$$D = \{x|f(x) < +\infty\},$$

and int$D$ and bd$D$ denote the interior and boundary of the set D. A closed proper convex function f is *essentially strictly convex* if f is strictly convex on every convex subset of dom ∂f.

The closed proper convex function f is essentially smooth if and only if the subdifferential $\partial f(x)$ is empty for $x \in$ bd$D$ and is $\{\nabla f(x)\}$ for $x \in$ int$D$ (so f is differentiable on int$D$) if and only if the function f^* is essentially strictly convex.

Definition A.1 *A closed proper convex function f is said to be a* Legendre function *if it is both essentially smooth and essentially strictly convex.*

So f is Legendre if and only if its conjugate function is Legendre, in which case the gradient operator ∇f is a topological isomorphism with

∇f^* as its inverse. The gradient operator ∇f maps int dom f onto int dom f^*. If int dom $f^* = \mathbb{R}^J$ then the range of ∇f is $\mathbb{R}^J$ and the equation $\nabla f(x) = y$ can be solved for every $y \in \mathbb{R}^J$. In order for int dom $f^* = \mathbb{R}^J$ it is necessary and sufficient that the Legendre function f be *super-coercive*, that is,

$$\lim_{||x||_2 \to +\infty} \frac{f(x)}{||x||_2} = +\infty. \tag{A.1}$$

If the effective domain of f is bounded, then f is super-coercive and its gradient operator is a mapping onto the space $\mathbb{R}^J$.

A.2 Bregman Projections onto Closed Convex Sets

Let f be a closed proper convex function that is differentiable on the nonempty set int$D$. The corresponding *Bregman distance* $D_f(x, z)$ is defined for $x \in \mathbb{R}^J$ and $z \in \text{int}D$ by

$$D_f(x, z) = f(x) - f(z) - \langle \nabla f(z), x - z \rangle. \tag{A.2}$$

Note that $D_f(x, z) \geq 0$ always and that $D_f(x, z) = +\infty$ is possible. If f is essentially strictly convex then $D_f(x, z) = 0$ implies that $x = z$.

Let K be a nonempty closed convex set with $K \cap \text{int}D \neq \emptyset$. Pick $z \in$ int$D$. The *Bregman projection* of z onto K, with respect to f, is

$$P_K^f(z) = \text{argmin}_{x \in K \cap D} D_f(x, z). \tag{A.3}$$

If f is essentially strictly convex, then $P_K^f(z)$ exists. If f is strictly convex on D then $P_K^f(z)$ is unique. If f is Legendre, then $P_K^f(z)$ is uniquely defined and is in int$D$; this last condition is sometimes called *zone consistency*.

Example: Let $J = 2$ and $f(x)$ be the function that is equal to one-half the norm squared on D, the nonnegative quadrant, $+\infty$ elsewhere. Let K be the set $K = \{(x_1, x_2) | x_1 + x_2 = 1\}$. The Bregman projection of $(2, 1)$ onto K is $(1, 0)$, which is not in int$D$. The function f is not essentially smooth, although it is essentially strictly convex. Its conjugate is the function f^* that is equal to one-half the norm squared on D and equal to zero elsewhere; it is essentially smooth, but not essentially strictly convex.

If f is Legendre, then $P_K^f(z)$ is the unique member of $K \cap \text{int}D$ satisfying the inequality

$$\langle \nabla f(P_K^f(z)) - \nabla f(z), P_K^f(z) - c \rangle \geq 0, \tag{A.4}$$

for all $c \in K$. From this we obtain the *Bregman Inequality*:

$$D_f(c,z) \geq D_f(c, P_K^f(z)) + D_f(P_K^f(z), z), \tag{A.5}$$

for all $c \in K$.

A.3 Bregman-Legendre Functions

Following Bauschke and Borwein [13], we say that a Legendre function f is a *Bregman-Legendre* function if the following properties hold:

B1: for x in D and any $a > 0$ the set $\{z | D_f(x,z) \leq a\}$ is bounded.
B2: if x is in D but not in int$D$, for each positive integer n, y^n is in int$D$ with $y^n \to y \in \text{bd}D$ and if $\{D_f(x,y^n)\}$ remains bounded, then $D_f(y,y^n) \to 0$, so that $y \in D$.
B3: if x^n and y^n are in int$D$, with $x^n \to x$ and $y^n \to y$, where x and y are in D but not in int$D$, and if $D_f(x^n, y^n) \to 0$ then $x = y$.

Bauschke and Borwein then prove that Bregman's SGP method converges to a member of K provided that one of the following holds: 1) f is Bregman-Legendre; 2) $K \cap \text{int}D \neq \emptyset$ and dom f^* is open; or 3) dom f and dom f^* are both open.

The Bregman functions form a class closely related to the Bregman-Legendre functions. For details see [39].

A.3.1 Useful Results about Bregman-Legendre Functions

The following results are proved in somewhat more generality in [13].
R1: If $y^n \in$ int dom f and $y^n \to y \in$ int dom f, then $D_f(y, y^n) \to 0$.
R2: If x and $y^n \in$ int dom f and $y^n \to y \in$ bd dom f, then $D_f(x, y^n) \to +\infty$.
R3: If $x^n \in D$, $x^n \to x \in D$, $y^n \in$ int D, $y^n \to y \in D$, $\{x,y\} \cap$ int $D \neq \emptyset$ and $D_f(x^n, y^n) \to 0$, then $x = y$ and $y \in$ int D.
R4: If x and y are in D, but are not in int D, $y^n \in$ int D, $y^n \to y$ and $D_f(x, y^n) \to 0$, then $x = y$.
As a consequence of these results we have the following.
R5: If $\{D_f(x, y^n)\} \to 0$, for $y^n \in$ int D and $x \in \mathbb{R}^J$, then $\{y^n\} \to x$.

Proof of R5: Since $\{D_f(x, y^n)\}$ is eventually finite, we have $x \in D$. By Property B1 above it follows that the sequence $\{y^n\}$ is bounded; without loss of generality, we assume that $\{y^n\} \to y$, for some $y \in \overline{D}$. If x is in int

D, then, by result R2 above, we know that y is also in int D. Applying result R3, with $x^n = x$, for all n, we conclude that $x = y$. If, on the other hand, x is in D, but not in int D, then y is in D, by result R2. There are two cases to consider: (1) y is in int D; (2) y is not in int D. In case (1) we have $D_f(x, y^n) \rightarrow D_f(x, y) = 0$, from which it follows that $x = y$. In case (2) we apply result R4 to conclude that $x = y$. ∎

Bibliography

[1] Anderson, A. and Kak, A. (1984) "Simultaneous algebraic reconstruction technique (SART): a superior implementation of the ART algorithm." *Ultrasonic Imaging*, **6** pp. 81–94.

[2] Attouch, H., Briceño-Arias, L.M., and Combettes, P. (2010) "A parallel splitting method for coupled monotone inclusions." *SIAM J. Control Optim.*, **48**, pp. 3246–3270.

[3] Attouch, H. (1984) *Variational Convergence for Functions and Operators*. Boston: Pitman Advanced Publishing Program.

[4] Attouch, H., and Wets, R. (1989) "Epigraphical analysis." *Ann. Inst. Poincare: Anal. Nonlineaire*, **6**.

[5] Aubin, J.-P., (1993) *Optima and Equilibria: An Introduction to Nonlinear Analysis*. Berlin: Springer-Verlag.

[6] Aubin, J.-P., and Ekeland, I. (1984) *Applied Nonlinear Analysis*. New York: Wiley.

[7] Auslander, A., and Teboulle, M. (2006) "Interior gradient and proximal methods for convex and conic optimization." *SIAM Journal on Optimization*, **16(3)**, pp. 697–725.

[8] Axelsson, O. (1994) *Iterative Solution Methods*. Cambridge, UK: Cambridge University Press.

[9] Baillon, J.-B., Bruck, R.E., and Reich, S. (1978) "On the asymptotic behavior of nonexpansive mappings and semigroups in Banach spaces." *Houston Journal of Mathematics*, **4**, pp. 1–9.

[10] Bauschke, H. (1996) "The approximation of fixed points of compositions of nonexpansive mappings in Hilbert space." *Journal of Mathematical Analysis and Applications*, **202**, pp. 150–159.

[11] Bauschke, H., and Borwein, J. (1993) "On the convergence of von Neumann's alternating projection algorithm for two sets." *Set-Valued Analysis*, **1**, pp. 185–212.

[12] Bauschke, H., and Borwein, J. (1996) "On projection algorithms for solving convex feasibility problems." *SIAM Review*, **38(3)**, pp. 367–426.

[13] Bauschke, H., and Borwein, J. (1997) "Legendre functions and the method of random Bregman projections." *Journal of Convex Analysis*, **4**, pp. 27–67.

[14] Bauschke, H., and Borwein, J. (2001) "Joint and separate convexity of the Bregman distance." In [38], pp. 23–36.

[15] Bauschke, H., and Combettes, P. (2001) "A weak-to-strong convergence principle for Fejér monotone methods in Hilbert spaces." *Mathematics of Operations Research*, **26**, pp. 248–264.

[16] Bauschke, H., and Combettes, P. (2003) "Iterating Bregman retractions." *SIAM Journal on Optimization*, **13**, pp. 1159–1173.

[17] Bauschke, H., Combettes, P., and Noll, D. (2006) "Joint minimization with alternating Bregman proximity operators." *Pacific Journal of Optimization*, **2**, pp. 401–424.

[18] Bauschke, H., and Combettes, P. (2011) *Convex Analysis and Monotone Operator Theory in Hilbert Spaces*, Canadian Mathematical Society Books in Mathematics. New York: Springer.

[19] Bauschke, H., and Lewis, A. (2000) "Dykstra's algorithm with Bregman projections: a convergence proof." *Optimization*, **48**, pp. 409–427.

[20] Bauschke, H., Burachik, R., Combettes, P., Elser, V., Luke, D., and Wolkowitz, H., eds. (2011) *Fixed-Point Algorithms for Inverse Problems in Science and Engineering*. New York: Springer-Verlag.

[21] Becker, M., Yang, I., and Lange, K. (1997) "EM algorithms without missing data." *Stat. Methods Med. Res.*, **6**, pp. 38–54.

[22] Berinde, V. (2007) *Iterative Approximation of Fixed Points*. Berlin: Springer-Verlag.

[23] Bertero, M., and Boccacci, P. (1998) *Introduction to Inverse Problems in Imaging*. Bristol, UK: Institute of Physics Publishing.

[24] Bertsekas, D.P. (1997) "A new class of incremental gradient methods for least squares problems." *SIAM J. Optim.*, **7**, pp. 913–926.

[25] Bertsekas, D., and Tsitsiklis, J. (1989) *Parallel and Distributed Computation: Numerical Methods*. New Jersey: Prentice-Hall.

[26] Bertsekas, D. (2003) *Convex Analysis and Optimization.* Nashua, NH: Athena Scientific.

[27] Borwein, J. and Lewis, A. (2000) *Convex Analysis and Nonlinear Optimization*, Canadian Mathematical Society Books in Mathematics. New York: Springer-Verlag.

[28] Boyd, S., and Vandenberghe, L. (2004) *Convex Optimization.* Cambridge, UK: Cambridge University Press.

[29] Boyles, R. (1983) "On the convergence of the EM algorithm." *J. Roy. Statist. Soc. B*, **45** pp. 47–50.

[30] Brauer, A. (1946) "Characteristic roots of a matrix." *Duke Mathematics Journal*, **13**, pp. 387–395.

[31] Bregman, L.M. (1967) "The relaxation method of finding the common point of convex sets and its application to the solution of problems in convex programming." *USSR Computational Mathematics and Mathematical Physics*, **7**, pp. 200–217.

[32] Bregman, L., Censor, Y., and Reich, S. (1999) "Dykstra's algorithm as the nonlinear extension of Bregman's optimization method." *Journal of Convex Analysis*, **6(2)**, pp. 319–333.

[33] Browne, E. (1930) "The characteristic roots of a matrix." *Bulletin of the American Mathematical Society*, **36**, pp. 705–710.

[34] Browne, J. and A. DePierro, A. (1996) "A row-action alternative to the EM algorithm for maximizing likelihoods in emission tomography." *IEEE Trans. Med. Imag.*, **15**, pp. 687–699.

[35] Bruck, R.E., and Reich, S. (1977) "Nonexpansive projections and resolvents of accretive operators in Banach spaces." *Houston Journal of Mathematics*, **3**, pp. 459–470.

[36] Bruckstein, A., Donoho, D., and Elad, M. (2009) "From sparse solutions of systems of equations to sparse modeling of signals and images." *SIAM Review*, **51(1)**, pp. 34–81.

[37] Burden, R.L., and Faires, J.D. (1993) *Numerical Analysis.* Boston: PWS-Kent.

[38] Butnariu, D., Censor, Y., and Reich, S., eds. (2001) *Inherently Parallel Algorithms in Feasibility and Optimization and Their Applications*, Studies in Computational Mathematics 8. Amsterdam: Elsevier Publ.

[39] Butnariu, D., Byrne, C., and Censor, Y. (2003) "Redundant axioms in the definition of Bregman functions." *Journal of Convex Analysis*, **10**, pp. 245–254.

[40] Byrne, C. and Fitzgerald, R. (1979) "A unifying model for spectrum estimation." In *Proceedings of the RADC Workshop on Spectrum Estimation*, Griffiss AFB, Rome, NY, October.

[41] Byrne, C. and Fitzgerald, R. (1982) "Reconstruction from partial information, with applications to tomography." *SIAM J. Applied Math.*, **42(4)**, pp. 933–940.

[42] Byrne, C., Fitzgerald, R., Fiddy, M., Hall, T., and Darling, A. (1983) "Image restoration and resolution enhancement." *J. Opt. Soc. Amer.*, **73**, pp. 1481–1487.

[43] Byrne, C. and Fitzgerald, R. (1984) "Spectral estimators that extend the maximum entropy and maximum likelihood methods." *SIAM J. Applied Math.*, **44(2)**, pp. 425–442.

[44] Byrne, C., Levine, B.M., and Dainty, J.C. (1984) "Stable estimation of the probability density function of intensity from photon frequency counts." *JOSA Communications*, **1(11)**, pp. 1132–1135.

[45] Byrne, C. and Fiddy, M. (1987) "Estimation of continuous object distributions from Fourier magnitude measurements." *JOSA A*, **4**, pp. 412–417.

[46] Byrne, C. and Fiddy, M. (1988) "Images as power spectra; reconstruction as Wiener filter approximation." *Inverse Problems*, **4**, pp. 399–409.

[47] Byrne, C. (1993) "Iterative image reconstruction algorithms based on cross-entropy minimization." *IEEE Transactions on Image Processing*, **IP-2**, pp. 96–103.

[48] Byrne, C. (1995) "Erratum and addendum to 'Iterative image reconstruction algorithms based on cross-entropy minimization'." *IEEE Transactions on Image Processing*, **IP-4**, pp. 225–226.

[49] Byrne, C. (1996) "Iterative reconstruction algorithms based on cross-entropy minimization." In *Image Models (and Their Speech Model Cousins)*, IMA Volumes in Mathematics and Its Applications 80, S.E. Levinson and L. Shepp, eds., pp. 1–11. New York: Springer-Verlag.

[50] Byrne, C. (1996) "Block-iterative methods for image reconstruction from projections." *IEEE Transactions on Image Processing*, **IP-5**, pp. 792–794.

[51] Byrne, C. (1997) "Convergent block-iterative algorithms for image reconstruction from inconsistent data." *IEEE Transactions on Image Processing*, **IP-6**, pp. 1296–1304.

[52] Byrne, C. (1998) "Accelerating the EMML algorithm and related iterative algorithms by rescaled block-iterative (RBI) methods." *IEEE Transactions on Image Processing*, **IP-7**, pp. 100–109.

[53] Byrne, C. (1998) "Iterative algorithms for deblurring and deconvolution with constraints." *Inverse Problems*, **14**, pp. 1455–1467.

[54] Byrne, C. (2000) "Block-iterative interior point optimization methods for image reconstruction from limited data." *Inverse Problems*, **16**, pp. 1405–1419.

[55] Byrne, C. (2001) "Bregman-Legendre multi-distance projection algorithms for convex feasibility and optimization." In [38], pp. 87–100.

[56] Byrne, C. (2001) "Likelihood maximization for list-mode emission tomographic image reconstruction." *IEEE Transactions on Medical Imaging*, **20(10)**, pp. 1084–1092.

[57] Byrne, C., and Censor, Y. (2001) "Proximity function minimization using multiple Bregman projections, with applications to split feasibility and Kullback-Leibler distance minimization." *Annals of Operations Research*, **105**, pp. 77–98.

[58] Byrne, C. (2002) "Iterative oblique projection onto convex sets and the split feasibility problem." *Inverse Problems*, **18**, pp. 441–453.

[59] Byrne, C. (2004) "A unified treatment of some iterative algorithms in signal processing and image reconstruction." *Inverse Problems*, **20**, pp. 103–120.

[60] Byrne, C. (2005) "Choosing parameters in block-iterative or ordered-subset reconstruction algorithms." *IEEE Transactions on Image Processing*, **14(3)**, pp. 321–327.

[61] Byrne, C. (2005) *Signal Processing: A Mathematical Approach.* Wellesley, MA: A K Peters.

[62] Byrne, C. (2007) *Applied Iterative Methods.* Wellesley, MA: A K Peters.

[63] Byrne, C. (2008) "Sequential unconstrained minimization algorithms for constrained optimization." *Inverse Problems*, **24(1)**, article no. 015013.

[64] Byrne, C. (2009) “Block-iterative algorithms.” *International Transactions in Operations Research*, **16(4)**, pp. 427–463.

[65] Byrne, C. (2009) “Bounds on the largest singular value of a matrix and the convergence of simultaneous and block-iterative algorithms for sparse linear systems.” *International Transactions in Operations Research*, **16(4)**, pp. 465–479.

[66] Byrne, C. (2013) “Alternating minimization as sequential unconstrained minimization: a survey.” *Journal of Optimization Theory and Applications*, electronic **154(3)**, DOI 10.1007/s1090134-2, 2012, and hardcopy **156(3)**, February, 2013, pp. 554–566.

[67] Byrne, C. (2013) “An elementary proof of convergence of the forward-backward splitting algorithm.” To appear in the *Journal of Nonlinear and Convex Analysis*.

[68] Byrne, C., and Eggermont, P. (2011) “EM Algorithms.” In *Handbook of Mathematical Methods in Imaging*, Otmar Scherzer, ed., pp. 271–344. New York: Springer-Science.

[69] Byrne, C., Censor, Y., A. Gibali, A., and Reich, S. (2012) “The split common null point problem.” *Journal of Nonlinear and Convex Analysis*, **13**, pp. 759–775.

[70] Byrne, C., and Ward, S. (2005) “Estimating the largest singular value of a sparse matrix.” Unpublished notes.

[71] Byrne, C., and Wells, D. (1983) “Limit of continuous and discrete finite-band Gerchberg iterative spectrum extrapolation.” *Optics Letters*, **8(10)**, pp. 526–527.

[72] Byrne, C., and Wells, D. (1985) “Optimality of certain iterative and non-iterative data extrapolation procedures.” *Journal of Mathematical Analysis and Applications*, **111(1)**, pp. 26–34.

[73] Candès, E., Romberg, J., and Tao, T. (2006) “Robust uncertainty principles: exact signal reconstruction from highly incomplete frequency information.” *IEEE Transactions on Information Theory*, **52(2)**, pp. 489–509.

[74] Candès, E., and Romberg, J. (2007) “Sparsity and incoherence in compressive sampling.” *Inverse Problems*, **23(3)**, pp. 969–985.

[75] Candès, E., Wakin, M., and Boyd, S. (2007) “Enhancing sparsity by reweighted l_1 minimization.” Preprint available at http://www.acm.caltech.edu/~emmanuel/publications.html.

[76] Cegielski, A. (2010) "Generalized relaxations of nonexpansive operators and convex feasibility problems." *Contemp. Math.*, **513**, pp. 111–123.

[77] Cegielski, A. (2012) *Iterative Methods for Fixed Point Problems in Hilbert Space*, Lecture Notes in Mathematics 2057. Heidelberg: Springer.

[78] Cegielski, A., and Censor, Y. (2011) "Opial-type theorems and the common fixed-point problem." In *Fixed-Point Algorithms for Inverse Problems in Science and Engineering*, Springer Optimization and Its Applications 49, H. Bauschke, R. Burachik, P. Combettes, V. Elser, D. Luke, and H. Wolkowicz, eds., pp. 155–183. New York: Springer.

[79] Cegielski, A., and Zalas, R. (2013) "Methods for the variational inequality problem over the intersection of fixed points of quasi-nonexpansive operators." *Numer. Funct. Anal. Optimiz.*, **34**, pp. 255–283.

[80] Censor, Y. (1981) "Row-action methods for huge and sparse systems and their applications." *SIAM Review*, **23**, pp. 444–464.

[81] Censor, Y., Bortfeld, T., Martin, B., and Trofimov, A. (2006) "A unified approach for inversion problems in intensity-modulated radiation therapy." *Physics in Medicine and Biology*, **51**, pp. 2353–2365.

[82] Censor, Y., Eggermont, P.P.B., and Gordon, D. (1983) "Strong underrelaxation in Kaczmarz's method for inconsistent systems." *Numerische Mathematik*, **41**, pp. 83–92.

[83] Censor, Y., and Elfving, T. (1994) "A multi-projection algorithm using Bregman projections in a product space." *Numerical Algorithms*, **8**, pp. 221–239.

[84] Censor, Y., Elfving, T., Herman, G.T., and Nikazad, T. (2008) "On diagonally-relaxed orthogonal projection methods." *SIAM Journal on Scientific Computation*, **30(1)**, pp. 473–504.

[85] Censor, Y., Elfving, T., Kopf, N., and Bortfeld, T. (2005) "The multiple-sets split feasibility problem and its application for inverse problems." *Inverse Problems*, **21**, pp. 2071–2084.

[86] Censor, Y., Gibali, A., and Reich, S. (2011) "The subgradient extragradient method for solving variational inequalities in Hilbert space." *Journal of Optimization Theory and Applications*, **148**, pp. 318–335.

[87] Censor, Y., Gibali, A., and Reich, S. (2012) "Algorithms for the split variational inequality problem." *Numerical Algorithms*, **59**, pp. 301–323.

[88] Censor, Y., Gordon, D., and Gordon, R. (2001) "Component averaging: an efficient iterative parallel algorithm for large and sparse unstructured problems." *Parallel Computing*, **27**, pp. 777–808.

[89] Censor, Y., Gordon, D., and Gordon, R. (2001) "BICAV: A block-iterative, parallel algorithm for sparse systems with pixel-related weighting." *IEEE Transactions on Medical Imaging*, **20**, pp. 1050–1060.

[90] Censor, Y., Iusem, A., and Zenios, S. (1998) "An interior point method with Bregman functions for the variational inequality problem with paramonotone operators." *Mathematical Programming*, **81**, pp. 373–400.

[91] Censor, Y., and Reich, S. (1996) "Iterations of paracontractions and firmly nonexpansive operators with applications to feasibility and optimization." *Optimization*, **37**, pp. 323–339.

[92] Censor, Y., and Reich, S. (1998) "The Dykstra algorithm for Bregman projections." *Communications in Applied Analysis*, **2**, pp. 323–339.

[93] Censor, Y., and Segman, J. (1987) "On block-iterative maximization." *J. of Information and Optimization Sciences*, **8**, pp. 275–291.

[94] Censor, Y., and Zenios, S.A. (1992) "Proximal minimization algorithm with D-functions." *Journal of Optimization Theory and Applications*, **73(3)**, pp. 451–464.

[95] Censor, Y., and Zenios, S.A. (1997) *Parallel Optimization: Theory, Algorithms and Applications*. New York: Oxford University Press.

[96] Cheney, W., and Goldstein, A. (1959) "Proximity maps for convex sets." *Proc. Amer. Math. Soc.*, **10**, pp. 448–450.

[97] Chidume, Ch. (2009) *Geometric Properties of Banach Spaces and Nonlinear Iterations*. London: Springer.

[98] Cimmino, G. (1938) "Calcolo approssimato per soluzioni dei sistemi di equazioni lineari."*La Ricerca Scientifica XVI, Series II, Anno IX*, **1**, pp. 326–333.

[99] Combettes, P. (2000) "Fejér monotonicity in convex optimization."In *Encyclopedia of Optimization*, C.A. Floudas and P.M. Pardalos, eds. Boston: Kluwer Publ.

[100] Combettes, P. (2001) "Quasi-Fejérian analysis of some optimization algorithms." In *Inherently Parallel Algorithms in Feasibility and Optimization and Their Applications*, Studies in Computational Mathematics 8, D. Butnariu, Y. Censor and S. Reich, eds., pp. 87–100. Amsterdam: Elsevier Publ.

[101] Combettes, P., and Wajs, V. (2005) "Signal recovery by proximal forward-backward splitting." *Multiscale Modeling and Simulation*, **4(4)**, pp. 1168–1200.

[102] Conn, A., Scheinberg, K., and Vicente, L. (2009) *Introduction to Derivative-Free Optimization*, MPS-SIAM Series on Optimization. Philadelphia: Society for Industrial and Applied Mathematics.

[103] Csiszár, I. (1975) "I-divergence geometry of probability distributions and minimization problems." *The Annals of Probability*, **3(1)**, pp. 146–158.

[104] Csiszár, I. (1989) "A geometric interpretation of Darroch and Ratcliff's generalized iterative scaling." *The Annals of Statistics*, **17(3)**, pp. 1409–1413.

[105] Csiszár, I., and Tusnády, G. (1984) "Information geometry and alternating minimization procedures." *Statistics and Decisions*, **Supp. 1**, pp. 205–237.

[106] Darroch, J., and Ratcliff, D. (1972) "Generalized iterative scaling for log-linear models." *Annals of Mathematical Statistics*, **43**, pp. 1470–1480.

[107] Dax, A. (1990) "The convergence of linear stationary iterative processes for solving singular unstructured systems of linear equations." *SIAM Review*, **32**, pp. 611–635.

[108] Dempster, A.P., Laird, N.M., and Rubin, D.B. (1977) "Maximum likelihood from incomplete data via the EM algorithm." *Journal of the Royal Statistical Society, Series B*, **37**, pp. 1–38.

[109] De Pierro, A., and Iusem, A. (1990) "On the asymptotic behavior of some alternate smoothing series expansion iterative methods." *Linear Algebra and Its Applications*, **130**, pp. 3–24.

[110] Deutsch, F., and Yamada, I. (1998) "Minimizing certain convex functions over the intersection of the fixed point sets of non-expansive mappings." *Numerical Functional Analysis and Optimization*, **19**, pp. 33–56.

[111] Donoho, D. (2006) "Compressed sampling." *IEEE Transactions on Information Theory*, **52(4)**, pp. 1289–1306. (Download preprints at http://www.stat.stanford.edu/~donoho/Reports).

[112] Duda, R., Hart, P., and Stork, D. (2001) *Pattern Classification*. New York: Wiley.

[113] Duffin, R., Peterson, E., and Zener, C. (1967) *Geometric Programming: Theory and Applications.* New York: Wiley.

[114] Dugundji, J. (1970) *Topology.* Boston: Allyn and Bacon, Inc.

[115] Dykstra, R. (1983) "An algorithm for restricted least squares regression." *J. Amer. Statist. Assoc.*, **78(384)**, pp. 837–842.

[116] Eggermont, P.P.B., Herman, G.T., and Lent, A. (1981) "Iterative algorithms for large partitioned linear systems, with applications to image reconstruction."*Linear Algebra and Its Applications*, **40**, pp. 37–67.

[117] Eggermont, P., and LaRiccia, V. (2001) *Maximum Penalized Likelihood Estimation.* New York: Springer.

[118] Elsner, L., Koltracht, L., and Neumann, M. (1992) "Convergence of sequential and asynchronous nonlinear paracontractions." *Numerische Mathematik*, **62**, pp. 305–319.

[119] Everitt, B., and Hand, D. (1981) *Finite Mixture Distributions.* London: Chapman and Hall.

[120] Facchinei, F., and Pang, J.S. (2003) *Finite Dimensional Variational Inequalities and Complementarity Problems, Volumes I and II.* New York: Springer Verlag.

[121] Fang, S-C., and Puthenpura, S. (1993) *Linear Optimization and Extensions: Theory and Algorithms.* New Jersey: Prentice-Hall.

[122] Farkas, J. (1902) "Über die Theorie der einfachen Ungleichungen." *J. Reine Angew. Math.*, **124**, pp. 1–24.

[123] Farncombe, T. (2000) "Functional dynamic SPECT imaging using a single slow camera rotation." Ph.D. thesis, Dept. of Physics, University of British Columbia.

[124] Farnell, A.B. (1944) "Limits for the characteristic roots of a matrix." *Bulletin of the American Mathematical Society*, **50**, pp. 789–794.

[125] Fessler, J., Ficaro, E., Clinthorne, N., and Lange, K. (1997) "Grouped-coordinate ascent algorithms for penalized-likelihood transmission image reconstruction." *IEEE Trans. Med. Imag.*, **16(2)**, pp. 166–175.

[126] Fiacco, A., and McCormick, G. (1990) *Nonlinear Programming: Sequential Unconstrained Minimization Techniques*, Classics in Mathematics (reissue). Philadelphia, PA: SIAM.

[127] Fiddy, M. (2008) *Private communication.*

[128] Geman, S., and Geman, D. (1984) "Stochastic relaxation, Gibbs distributions and the Bayesian restoration of images." *IEEE Transactions on Pattern Analysis and Machine Intelligence*, **PAMI-6**, pp. 721–741.

[129] Gill, P., Murray, W., and Wright, M. (1981) *Practical Optimization.* San Diego: Academic Press.

[130] Gill, P., Murray, W., Saunders, M., Tomlin, J., and Wright, M. (1986) "On projected Newton barrier methods for linear programming and an equivalence to Karmarkar's projective method." *Mathematical Programming*, **36**, pp. 183–209.

[131] Goebel, K., and Reich, S. (1984) *Uniform Convexity, Hyperbolic Geometry, and Nonexpansive Mappings.* New York: Dekker.

[132] Goldstein, S., and Osher, S. (2008) "The split Bregman algorithm for L^1 regularized problems. " UCLA CAM Report 08-29, UCLA, Los Angeles, CA.

[133] Golshtein, E., and Tretyakov, N. (1996) *Modified Lagrangians and Monotone Maps in Optimization.* New York: John Wiley and Sons, Inc.

[134] Gordon, R., Bender, R., and Herman, G.T. (1970) "Algebraic reconstruction techniques (ART) for three-dimensional electron microscopy and x-ray photography." *J. Theoret. Biol.*, **29**, pp. 471–481.

[135] Gordon, D., and Gordon, R.(2005) "Component-averaged row projections: A robust block-parallel scheme for sparse linear systems." *SIAM Journal on Scientific Computing*, **27**, pp. 1092–1117.

[136] Grcar, J. (2011) "John von Neumann's analysis of Gaussian elimination and the origins of modern numerical analysis." *SIAM Review*, **53(4)**, pp. 607–682.

[137] Gubin, L.G., Polyak, B.T. and Raik, E.V. (1967) "The method of projections for finding the common point of convex sets." *USSR Computational Mathematics and Mathematical Physics*, **7**, pp. 1–24.

[138] Hager, W. (1988) *Applied Numerical Linear Algebra.* Englewood Cliffs, NJ: Prentice Hall.

[139] Hager, B., Clayton, R., Richards, M., Comer, R., and Dziewonsky, A. (1985) "Lower mantle heterogeneity, dynamic typography and the geoid." *Nature*, **313**, pp. 541–545.

[140] Herman, G.T. (1999) *Private communication.*

[141] Herman, G.T., and Meyer, L. (1993) "Algebraic reconstruction techniques can be made computationally efficient." *IEEE Transactions on Medical Imaging*, **12**, pp. 600–609.

[142] Hildreth, C. (1957) "A quadratic programming procedure." *Naval Research Logistics Quarterly*, **4**, pp. 79–85. Erratum, ibid., p. 361.

[143] Hiriart-Urruty, J.-B., and Lemaréchal, C. (2001) *Fundamentals of Convex Analysis*. Berlin: Springer.

[144] Hirstoaga, S.A. (2006) "Iterative selection methods for common fixed point problems." *J. Math. Anal. Appl.*, **324**, pp. 1020–1035.

[145] Hogg, R., McKean, J., and Craig, A. (2004) *Introduction to Mathematical Statistics*, 6th edition. Englewood Cliffs, NJ: Prentice Hall.

[146] Holte, S., Schmidlin, P., Linden, A., Rosenqvist, G. and Eriksson, L. (1990) "Iterative image reconstruction for positron emission tomography: a study of convergence and quantitation problems." *IEEE Transactions on Nuclear Science*, **37**, pp. 629–635.

[147] Jiang, M., and Wang, G. (2003) "Convergence studies on iterative algorithms for image reconstruction." *IEEE Transactions on Medical Imaging*, **22(5)**, pp. 569–579.

[148] Kaczmarz, S. (1937) "Angenäherte Auflösung von Systemen linearer Gleichungen." *Bulletin de l'Academie Polonaise des Sciences et Lettres*, **A35**, pp. 355–357.

[149] Karmarkar, N. (1984) "A new polynomial-time algorithm for linear programming." *Combinatorica*, **4**, pp. 373–395.

[150] Kelley, C.T. (1999) *Iterative Methods for Optimization*, Frontiers in Applied Mathematics. Philadelphia: SIAM Publications.

[151] Korpelevich, G. (1976) "The extragradient method for finding saddle points and other problems." *Ekonomika i Matematcheskie Metody* (in Russian), **12**, pp. 747–756.

[152] Krasnosel'skii, M. (1955) "Two observations on the method of sequential approximations." *Uspeki Mathematicheskoi Nauki* (in Russian), **10(1)**, pp. 123–127.

[153] Kullback, S. and Leibler, R. (1951) "On information and sufficiency." *Annals of Mathematical Statistics*, **22**, pp. 79–86.

[154] Landweber, L. (1951) "An iterative formula for Fredholm integral equations of the first kind." *Amer. J. of Math.*, **73**, pp. 615–624.

[155] Lange, K. and Carson, R. (1984) "EM reconstruction algorithms for emission and transmission tomography." *Journal of Computer Assisted Tomography*, **8**, pp. 306–316.

[156] Lange, K., Bahn, M. and Little, R. (1987) "A theoretical study of some maximum likelihood algorithms for emission and transmission tomography." *IEEE Trans. Med. Imag.*, **MI-6(2)**, pp. 106–114.

[157] Lange, K., Hunter, D., and Yang, I. (2000) "Optimization transfer using surrogate objective functions (with discussion)." *J. Comput. Graph. Statist.*, **9**, pp. 1–20.

[158] Leahy, R., and Byrne, C. (2000) "Guest editorial: Recent development in iterative image reconstruction for PET and SPECT." *IEEE Trans. Med. Imag.*, **19**, pp. 257–260.

[159] Lent, A., and Censor, Y. (1980) "Extensions of Hildreth's row-action method for quadratic programming." *SIAM Journal on Control and Optimization*, **18**, pp. 444–454.

[160] Levy, A. (2009) *The Basics of Practical Optimization.* Philadelphia: SIAM Publications.

[161] Lucet, Y. (2010) "What shape is your conjugate? A survey of computational convex analysis and its applications." *SIAM Review*, **52(3)**, pp. 505–542.

[162] Luenberger, D. (1969) *Optimization by Vector Space Methods.* New York: John Wiley and Sons, Inc.

[163] Luo, Z., Ma, W., So, A., Ye, Y., and Zhang, S. (2010) "Semidefinite relaxation of quadratic optimization problems." *IEEE Signal Processing Magazine*, **27(3)**, pp. 20–34.

[164] Mann, W. (1953) "Mean value methods in iteration." *Proc. Amer. Math. Soc.*, **4**, pp. 506–510.

[165] Marzetta, T. (2003) "Reflection coefficient (Schur parameter) representation for convex compact sets in the plane." *IEEE Transactions on Signal Processing*, **51(5)**, pp. 1196–1210.

[166] McKinnon, K. (1998) "Convergence of the Nelder-Mead simplex method to a non-stationary point." *SIAM Journal on Optimization*, **9(1)**, pp. 148–158.

[167] McLachlan, G.J., and Krishnan, T. (1997) *The EM Algorithm and Extensions.* New York: John Wiley and Sons, Inc.

[168] Metropolis, N., Rosenbluth, A., Rosenbluth, M., Teller, A., and Teller, E. (1953) "Equation of state calculations by fast computing machines." *J. Chem. Phys.*, **21**, pp. 1087–1091.

[169] Moreau, J.-J. (1962) "Fonctions convexes duales et points proximaux dans un espace hilbertien." *C.R. Acad. Sci. Paris Sér. A Math.*, **255**, pp. 2897–2899.

[170] Moreau, J.-J. (1963) "Propriétés des applications 'prox'." *C.R. Acad. Sci. Paris Sér. A Math.*, **256**, pp. 1069–1071.

[171] Moreau, J.-J. (1965) "Proximité et dualité dans un espace hilbertien." *Bull. Soc. Math. France*, **93**, pp. 273–299.

[172] Moudafi, A. (2011) "Split monotone variation inclusions." *Journal of Optimization Theory and Applications*, **150**, pp. 275–283.

[173] Narayanan, M., Byrne, C. and King, M. (2001) "An interior point iterative maximum-likelihood reconstruction algorithm incorporating upper and lower bounds with application to SPECT transmission imaging." *IEEE Transactions on Medical Imaging*, **TMI-20(4)**, pp. 342–353.

[174] Nash, S., and Sofer, A. (1996) *Linear and Nonlinear Programming*. New York: McGraw-Hill.

[175] Nelder, J., and Mead, R. (1965) "A simplex method for function minimization." *Computing Journal*, **7**, pp. 308–313.

[176] Nesterov, Y., and Nemirovski, A. (1994) *Interior-Point Polynomial Algorithms in Convex Programming.* Philadelphia, PA: SIAM Studies in Applied Mathematics.

[177] von Neumann, J., and Morgenstern, O. (1944) *Theory of Games and Economic Behavior.* Princeton, NJ: Princeton University Press.

[178] Noor, M.A. (1999) "Some algorithms for general monotone mixed variational inequalities." *Mathematical and Computer Modelling*, **29**, pp. 1–9.

[179] Noor, M.A. (2003) "Extragradient methods for pseudomonotone variational inequalities." *Journal of Optimization Theory and Applications*, **117(3)**, pp. 475–488.

[180] Noor, M.A. (2004) "Some developments in general variational inequalities." *Applied Mathematics and Computation*, **152**, pp. 199–277.

[181] Noor, M.A. (2010) "On an implicit method for nonconvex variational inequalities." *Journal of Optimization Theory and Applications*, **147**, pp. 411–417.

[182] Opial, Z. (1967) "Weak convergence of the sequence of successive approximations for nonexpansive mappings." *Bulletin of the American Mathematical Society*, **73**, pp. 591–597.

[183] Ortega, J., and Rheinboldt, W. (2000) *Iterative Solution of Nonlinear Equations in Several Variables*, Classics in Applied Mathematics 30. Philadelphia, PA: SIAM.

[184] Osher, S., Mao, Y., Dong, B., and Yin, W. (2008) "Fast linearized Bregman iterations for compressed sensing and sparse denoising." UCLA CAM Report 08-37.

[185] Papoulis, A. (1977) *Signal Analysis*. New York: McGraw-Hill.

[186] Peressini, A., Sullivan, F., and Uhl, J. (1988) *The Mathematics of Nonlinear Programming*. New York: Springer-Verlag.

[187] Redner, R., and Walker, H. (1984) "Mixture densities, maximum likelihood and the EM algorithm." *SIAM Review*, **26(2)**, pp. 195–239.

[188] Reich, S. (1979) "Weak convergence theorems for nonexpansive mappings in Banach spaces." *Journal of Mathematical Analysis and Applications*, **67**, pp. 274–276.

[189] Reich, S. (1980) "Strong convergence theorems for resolvents of accretive operators in Banach spaces." *Journal of Mathematical Analysis and Applications*, **75(1)**, pp. 287–292.

[190] Reich, S. (1996) "A weak convergence theorem for the alternating method with Bregman distances." In *Theory and Applications of Nonlinear Operators of Accretive and Monotone Type*, A.G. Kartsatos, ed., p. 313. New York: Dekker.

[191] Renegar, J. (2001) *A Mathematical View of Interior-Point Methods in Convex Optimization*, MPS-SIAM Series on Optimization. Philadelphia, PA: SIAM

[192] Rockafellar, R. (1970) *Convex Analysis*. Princeton, NJ: Princeton University Press.

[193] Rockmore, A., and Macovski, A. (1976) "A maximum likelihood approach to emission image reconstruction from projections." *IEEE Transactions on Nuclear Science*, **NS-23**, pp. 1428–1432.

[194] Saad, Y. (2003) *Iterative Methods for Sparse Linear Systems*, 2nd edition. Philadelphia: SIAM Publications.

[195] Schmidlin, P. (1972) "Iterative separation of sections in tomographic scintigrams." *Nuklearmedizin*, **11**, pp. 1–16.

[196] Shepp, L., and Vardi, Y. (1982) "Maximum likelihood reconstruction for emission tomography." *IEEE Transactions on Medical Imaging*, **MI-1**, pp. 113–122.

[197] Shieh, M., Byrne, C., and Fiddy, M. (2006) "Image reconstruction: a unifying model for resolution enhancement and data extrapolation: Tutorial." *Journal of the Optical Society of America, A*, **23(2)**, pp. 258–266.

[198] Shieh, M., Byrne, C., Testorf, M., and Fiddy, M. (2006) "Iterative image reconstruction using prior knowledge." *Journal of the Optical Society of America, A*, **23(6)**, pp. 1292–1300.

[199] Shieh, M., and Byrne, C. (2006) "Image reconstruction from limited Fourier data." *Journal of the Optical Society of America, A*, **23(11)**, pp. 2732–2736.

[200] Stark, H., and Yang, Y. (1998) *Vector Space Projections. A Numerical Approach to Signal and Image processing, Neural Nets and Optics.* New York: John Wiley and Sons.

[201] Stark, H., and Woods, J. (2002) *Probability and Random Processes, with Applications to Signal Processing.* Upper Saddle River, NJ: Prentice-Hall.

[202] Tanabe, K. (1971) "Projection method for solving a singular system of linear equations and its applications."*Numer. Math.*, **17**, pp. 203–214.

[203] Teboulle, M. (1992) "Entropic proximal mappings with applications to nonlinear programming." *Mathematics of Operations Research*, **17(3)**, pp. 670–690.

[204] van der Sluis, A. (1969) "Condition numbers and equilibration of matrices." *Numer. Math.*, **14**, pp. 14–23.

[205] van der Sluis, A., and van der Vorst, H.A. (1990) "SIRT- and CG-type methods for the iterative solution of sparse linear least-squares problems." *Linear Algebra and Its Applications*, **130**, pp. 257–302.

[206] Vardi, Y., Shepp, L.A. and Kaufman, L. (1985) "A statistical model for positron emission tomography."*Journal of the American Statistical Association*, **80**, pp. 8–20.

[207] von Neumann, J., and Goldstine, H. H. (1947) "Numerical inverting of matrices of high order." *Bulletin of the American Mathematical Society*, **53**, pp. 1021–1099.

[208] Wright, M. (2005) "The interior-point revolution in optimization: history, recent developments, and lasting consequences." *Bulletin (New Series) of the American Mathematical Society*, **42(1)**, pp. 39–56.

[209] Wright, M. (2009) "The dual flow between linear algebra and optimization." View-graphs of talk given at the History of Numerical Linear Algebra Minisymposium - Part II, SIAM Conference on Applied Linear Algebra, Monterey, CA, October 28, 2009.

[210] Wu, C.F.J. (1983) "On the convergence properties of the EM algorithm." *Annals of Stat.*, **11**, pp. 95–103.

[211] Yang, Q. (2004) "The relaxed CQ algorithm solving the split feasibility problem." *Inverse Problems*, **20**, pp. 1261–1266.

[212] Yamada, I. (2001) "The hybrid steepest descent method for the variational inequality problem over the intersection of fixed points of nonexpansive mappings." In [38], pp. 473–504.

[213] Yamada, I., and Ogura, N. (2004) "Hybrid steepest descent method for the variational inequality problem over the fixed point set of certain quasi-nonexpansive mappings." Optimiz., **25**, pp. 619–655.

Index